HOBBY HYDROPONICS

Second Edition

HOBBY HYDROPONICS

Second Edition

Howard M. Resh

CRC Press
Taylor & Francis Group
Boca Raton London New York

CRC Press is an imprint of the
Taylor & Francis Group, an **informa** business

CRC Press
Taylor & Francis Group
6000 Broken Sound Parkway NW, Suite 300
Boca Raton, FL 33487-2742

Printed in the United States of America ·
Version Date: 2012918

International Standard Book Number: 978-1-4665-6941-6 (Paperback)

Library of Congress Cataloging-in-Publication Data

Resh, Howard M.
   Hobby hydroponics / Howard M. Resh. -- 2nd ed.
     p. cm.
   Includes bibliographical references and index.
   ISBN 978-1-4665-6941-6 (pbk.   )
   1. Hydroponics. I. Title.

SB126.5.R46 2013
631.5'85--dc23   2012035757

Visit the Taylor & Francis Web site at
http://www.taylorandfrancis.com

and the CRC Press Web site at
http://www.crcpress.com

Contents

List of Figures

List of Tables

Preface

This book was originally written in 2003. Since that time there has been a huge advancement in technology for small-scale hydroponic units. Many new products and systems have been developed. The purpose of this updated edition is to include these new concepts and products built upon the information presented in the first edition. Many new photos and drawings are included to assist in understanding the products available at this time.

The objective of the book is to provide the reader with information on some of the typical hobby hydroponic units available in the marketplace. It is not my intention to favor some over others but to give the reader sufficient background in the type of units to purchase for the crops to be grown. The prices for these units vary greatly, so one can acquire items that are within one's budget. Along with the hydroponic units are needed accessories and supplies to make them grow plants successfully, so these will also be discussed. Some accessories include lights, carbon dioxide generating systems, and pH and electric conductivity (EC) monitoring instruments. In addition, supplies that are part of growing include nutrients, pest control, propagation cubes, blocks, trays, and growing slabs.

The book also instructs you how to start your plants, recommends crop varieties, describes seed sources, outlines training techniques for the various crops, and suggests what environmental conditions are best to assist you in successful growing. In addition, to "plug-and-play" units that you can purchase, I will instruct you on how to construct some systems for yourself, if you wish. The most common growing systems, such as water culture, nutrient film technique (NFT), perlite, coco coir, peat mixes and others, are described along with the hobby units using such cultures.

There is nothing mysterious about hydroponic growing; it is really following instructions and adhering to details in plant training, and monitoring the nutrient solution and the surrounding environment to obtain as close as possible to optimum conditions that make you successful.

Wishing you successful growing!

About the Author

Howard M. Resh (born January 11, 1941) is a recognized authority worldwide on hydroponics. His Web site www.howardresh.com presents information on hydroponic culture of various vegetable crops. In addition, he has written five books on hydroponic culture both for commercial growers and backyard hobbyists. While a graduate student at the University of British Columbia, in Vancouver, Canada, in 1971, he was asked by a private group to assist it in the construction of hydroponic greenhouses in the Vancouver area. He continued with outside work in greenhouses and soon was asked to conduct evening extension courses in hydroponics.

Upon graduation with his doctorate degree in horticulture in 1975, Resh became urban horticulturist for the Faculty of Plant Science at the University of British Columbia. He held that position for 3 years before the call of commercial hydroponics took him to many projects in countries including Venezuela, Taiwan, Saudi Arabia, the United States, and in 1999 to Anguilla, British West Indies, in the Eastern Caribbean, where he is still today.

While in the position of urban horticulturist, Resh taught courses in horticulture, hydroponics, plant propagation, and greenhouse design and production. While urban horticulturist and later general manager for a large plant nursery, Resh continued doing research and production consultation for a commercial hydroponic farm growing lettuce, watercress, and other vegetables in Venezuela. Later during 1995 to 1996, Resh became project manager for the Venezuelan farm to develop hydroponic culture of lettuce, watercress, peppers, tomatoes, and European cucumbers using a special medium of rice hulls and coco coir from local sources. He also designed and constructed a mung bean and alfalfa sprout facility to introduce sprouts into the local market.

In the late 1980s Resh worked with a company in Florida in the growing of lettuce in a floating raft culture system.

From 1990 to 1999, Resh worked as the technical director and project manager for hydroponic projects in the growing of watercress and herbs in California. He designed and constructed several 3-acre outdoor hydroponic watercress facilities using a unique nutrient film technique (NFT) system. These overcame production losses due to drought conditions in the area.

From there in mid-1999, Resh became the hydroponic greenhouse farm manager for the first hydroponic farm associated with a high-end resort, CuisinArt Resort & Spa, in Anguilla. The hydroponic farm is unique in being the only one in the world owned by a resort growing its own fresh salad crops and herbs exclusively for the resort. This farm has become a key component of the resort in attracting guests to experience home-grown vegetables, including tomatoes, cucumbers, peppers, lettuce, bok choy, and herbs. The resort, together with its hydroponic farm, has gained worldwide recognition as one of the leading hotels of the world.

Resh continues to do consulting on many unique hydroponic greenhouse operations, including Lufa Farms in Montreal, Canada. There he established the growing techniques and hydroponic systems for a rooftop hydroponic greenhouse in downtown Montreal. All vegetables are marketed through a community-supported agriculture (CSA) program.

1 Background

Whereas the commercialization of hobby or "popular" hydroponics became very active in the past 25 to 30 years, simple, self-designed systems were in use since the 1940s and 1950s. At that time the most common methods were gravel culture and water culture.

The bucket system of gravel culture was the most common, in which a small bed or bucket of gravel had a bucket attached to a hose joining them (Figure 1.1). A bucket was filled with nutrient solution and then supported above the growing bed to allow the nutrient solution to flow into the bed flooding it from the bottom upward. This was a subirrigation gravel culture system. Within a period of 5 minutes when the bucket completely drains into the substrate void spaces of the bed, it is lowered below the level of the bed and the nutrient solution drains back to the bucket. This is repeated by hand a number of times a day depending upon the water demand of the plants, which is determined by the temperature, light, and stage of plant growth. Although the system is very simple and works well, it requires that the hobbyist spend time during the day caring for the plants' irrigation needs.

A water culture system termed a "litter tray" system used a reservoir of nutrient solution in a rectangular tank with a tray of substrate located above the nutrient solution (Figure 1.2). The most common medium was excelsior wood fibers, wood shavings, sawdust, dry straw, rice hulls, or peanut hulls. The tray was 2 to 4 inches thick, constructed of wood with a wire mesh on the bottom to hold the substrate. Galvanized chicken wire of 1-inch diameter was coated with asphalt paint to prevent the release of zinc from its galvanized coating. The nutrient tank had a depth of 4 to 6 inches. The substrate would be watered when transplanting by hand for several days until the roots began growing into the nutrient solution below. Once the roots of the plant became established into the nutrient solution below, the solution level would be lowered gradually from 1 inch to 2 to 3 inches between the top of the solution and the base of the litter tray. This helped oxygenate the plant roots.

With the introduction of plastics, small pumps, timers, and drip irrigation supplies, these similar designs could be modified to operate automatically. For example, using the principle of the litter tray, construct the tray of plastic sitting on top of a rigid plastic or fiberglass nutrient reservoir. The growing tray contains 2 to 4 inches of a substrate such as lightweight

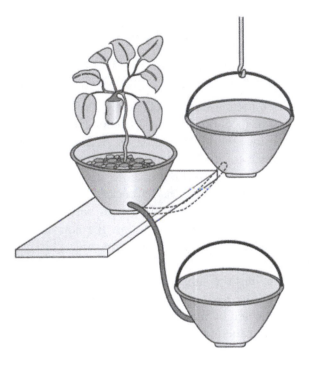

FIGURE 1.1 Bucket system of gravel culture. (Drawing courtesy of George Barile, Accurate Art, Inc., Holbrook, New York.)

volcanic rock, leca-clay pellets, haydite-porous shale rock, perlite, sawdust, bark, or a mixture of rice hulls, peat, or coco coir. Small holes of about ½-inch diameter are drilled in the bottom of the tray. A plastic screen is placed on the bottom of the tray to prevent the substrate from falling through into the nutrient tank below. A small fountain pump plugged into a simple household time clock will operate preset irrigation cycles. The pump is attached to a ½-inch diameter black polyethylene hose containing trickle tubes with irrigation emitters as it rises on top of the growing tray. The excess nutrient solution percolates through the perforated bottom of the growing tray back to the nutrient reservoir below.

One of the earlier automated hobby units was the "city green" hydro-ponicums. These units were available in the 1970s. They were the first commercial attempt at small hobby hydroponic systems. They were con-structed of molded plastic having a nutrient reservoir in the bottom and an upper growing tray. Expanded clay or volcanic cinder rock was the choice of growing medium. A fish-aquarium air pump was placed in one corner of the growing tray where it was attached to a small polyethylene tube that entered the nutrient tank below. This tube was inserted about 1½ inches into a slightly larger diameter tube allowing a small space between the walls of the two tubes (Figure 1.3). A pin held them together. As the

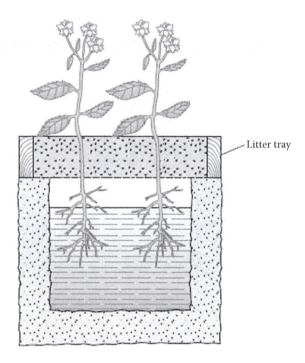

FIGURE 1.2   Cross-section of a typical water-culture bed. (Drawing courtesy of George Barile, Accurate Art, Inc., Holbrook, New York.)

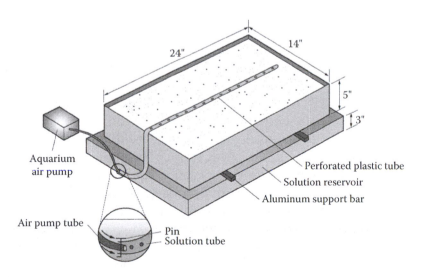

FIGURE 1.3   Components of an indoor unit. Air from an aquarium pump is used to move the nutrient solution up a tube to the growing tray. (Drawing courtesy of George Barile, Accurate Art, Inc., Holbrook, New York.)

FIGURE 1.4 Inverted bottle in a nursery tray system for growing herbs and lettuce.

aquarium pump forced air from the smaller tube into the larger one, the nutrient solution would be sucked into the tube together with the air bubbles. The solution with the air bubbles would rise in the larger tube to the surface of the growing tray where the tube was perforated to permit the nutrient solution to escape along with the air. The nutrient solution percolates through the medium and back to the reservoir underneath through the perforated bottom of the growing tray.

Perhaps one of the simplest hydroponic systems is the inverted bottle in a tray (Figure 1.4). For the solution tray use a plastic flat of 10½ × 21 inches that has no holes. Place a bedding tray of 24 or 36 compartments in the flat, but remove one corner of the filler tray to allow room for placement of a 1-gallon plastic jar. It must have a large plastic lid. Drill a ½-inch diameter hole in the middle of the large cap and glue a split cork ring of 3 inches in diameter on the cap. The small gap will allow the flow of nutrients from the bottle into the tray as the plants take up the solution. The bottle is inverted in the bottom of the tray. As solution flows to the plants, the level is maintained by air entering the bottle through the hole in the lid permitting a small amount of solution to flow from the bottle.

Use vermiculite or perlite as a substrate. You can seed directly into the medium. Water the seeds for several days until germination occurs before placing the inverted solution reservoir in the tray. You can cover the tray with plastic for several days until germination starts, then immediately remove the plastic or the seedlings will get long and leggy from excess heat. This tray is good for baby lettuce and herbs, or a combination

FIGURE 1.5 Small home unit using a perlite–vermiculite medium.

of lettuce, beets, herbs, upland cress, arugula, mustards, mizuna, orach, chard, and spinach to form a mesclun mix.

Over the past 30 years with increasing interest in hobby hydroponics, many small-scale units have been developed to meet market demand (Figure 1.5). Now it is very easy to visit hydroponic stores in person or online worldwide to purchase hobby units and all the supplies you will need to get started. A list of some of the common sites and publications is given in the last chapter.

The objective of this book is to make you aware of the types of hydroponic hobby units that are available on the market today and the supplies you need to get started. I shall describe in detail some of the units available and suggest which crops they are most suitable to growing. This information I hope will assist you in your decision to enter into hobby hydroponics.

Hobby hydroponics will provide you with the benefits of pleasure, rewarding products, clean products, and relaxation. You will achieve self-satisfaction by growing your own "garden fresh" salads. These salads will be nutritious and free of pesticides. Nutritional analyses of hydroponic tomatoes and peppers have demonstrated increases of up to 50% in vitamin and mineral content. These included vitamins A, B1 (thiamin), B2 (riboflavin), B3 (niacin), B6 (pyridoxine), C, and E. By using bioagents, beneficial insects, and natural pest control measures, your product will be free of synthetic pesticides. You will also feel relaxed and relieved from everyday working stress as you attend your plants in the hydroponic garden. It will allow your mind to escape from your daily concerns as you train and care for your plants. This is especially helpful during the dark days of winter as you are looking after your plants under supplementary lighting.

Another aspect of the hobby is that there are many Web sites, links, and e-mail addresses on the Internet to get assistance and new ideas for your hydroponic growing. You may ask questions or just wish to "chat" with other hobbyists growing hydroponically. There are hydroponic associations in most countries that have annual conferences and regular meetings in which you may participate to learn new things and meet new friends having similar interests as yourself. I describe some of these in the last chapter.

2 Starting Your Plants

SEEDS

The best method of starting your plants is from seed. The choice of variety of any plant is important. You should use those varieties that have been proven to grow best under hydroponic culture. Although all varieties will grow well hydroponically, some special greenhouse varieties developed for controlled environmental conditions, such as you will have in your house or a hobby greenhouse, grow faster and yield higher than conventional field varieties. I have listed in Table 2.1 some varieties of lettuce, arugula, herbs, bok choy, tomatoes, peppers, eggplants, and European cucumbers that I have found do well in hydroponic culture. This table is only a guideline, as there are many other varieties available that may produce well under your conditions. Feel free to test them yourself.

Base your choice of variety on the kind of product you want, your indoor conditions, and productivity. For example, you may wish a loose-leaf lettuce instead of a Bibb type, a cherry tomato not a beefsteak one; you may prefer a certain color of pepper and a particular fragrance of an herb. Your indoor conditions are important. For instance, during the cooler winter months choose varieties that will tolerate lower temperatures and light conditions. Seed catalogs will tell you which varieties do better under lower temperatures and light. Particular varieties will yield higher than others under these environmental conditions, so it is not always the best choice to look only at those highest yielding varieties as they may require much higher temperatures and light than you are able to provide.

SOWING OF SEEDS IN A MEDIUM

You must sow your seeds in some substrate. Some considerations of your choice of substrate include the plant, the hydroponic culture system, water retention, oxygenation, structural integrity, sterility, and ease of handling.

There are many different media that can be used to sow your seeds (Figure 2.1). A standard method used in raising bedding plants is that of plastic multipack trays in flats. The multipack trays come in many different compartments per tray. All fit into a standard 10½ × 21-inch flat. The most common ones for tomatoes and peppers would be the 36-compartment trays. For lettuce and herbs you could use the 72-compartment trays. It is best not to use these trays for cucumbers. With the trays you need a

7

TABLE 2.1
Plant Varieties and Sources

Plant	Variety	Source	Notes
Arugula	Roquette, Astro	Johnny's Selected Seeds	
Bok choy	Green Fortune	Ornamental Edibles	
European cucumber	Dominica, Marillo, Camaro	Paramount Seeds	Mildew resistant Partial mildew resistant
Mini (BA) cucumber	Manar, Jawell		Partial mildew resistant
Eggplants	Taurus	Paramount Seeds	Greenhouse staking
	Hansel, Fairy Tale	Johnny's Selected Seeds	Bush variety
Herbs	Sweet Italian Basil, Purple Basil, Chervil, Chives, Cilantro (Coriander), Dill-Fernleaf, Spearmint, Lavender, Oregano-Greek, Parsley-Curled, Parsley-Italian, Rosemary, Sage, Summer Savory, Sweet Marjoram, Thyme, Watercress	Stokes Seeds, Johnny's Selected Seeds, Richters	Use primed seeds for rosemary
Lettuce	Rex	Paramount Seeds	Bibb/Buttercrunch
	Vegas	Paramount Seeds	Bibb/Buttercrunch
	Charles	Paramount Seeds	Bibb/Buttercrunch
	Sylvesta	Johnny's Selected Seeds, Stokes Seeds	Bibb/Buttercrunch
	Skyphos	Johnny's Selected Seeds, Stokes Seeds	Bibb/Buttercrunch
	Helvius	Johnny's Selected Seeds, Stokes Seeds	Romaine/Cos
	Parris Island	Johnny's Selected Seeds, Stokes Seeds	Romaine/Cos

TABLE 2.1 (*Continued*)
Plant Varieties and Sources

Plant	Variety	Source	Notes
	Outredgeous	Johnny's Selected Seeds, Stokes Seeds	Romaine/Cos
	Lolla Rossa	Johnny's Selected Seeds, Stokes Seeds	Red Looseleaf
	Antago	Johnny's Selected Seeds, Stokes Seeds	Red Looseleaf
	Soltero	Johnny's Selected Seeds, Stokes Seeds	Red Looseleaf
	Malice	Johnny's Selected Seeds, Stokes Seeds	Green Looseleaf
	Ferrari	Johnny's Selected Seeds, Stokes Seeds	Red Oakleaf
	Oscarde	Johnny's Selected Seeds, Stokes Seeds	Red Oakleaf
	Navarra	Paramount Seeds	Red Oakleaf
	Cocarde	Paramount Seeds	Green Oakleaf
	Multy-Red/Green	Paramount Seeds	Leafy curled
Microgreens	Various	Johnny's Selected Seeds	Purchase seeds for varieties to give you the flavor that you want; see catalog
Peppers, staking bell type	Fantasy-Red, Paramo, Magno-Orange, Lesley, Cigales-Yellow	Paramount Seeds	Staking pepper
Bush Varieties	King Arthur-Red, Bianca-Ivory	Stokes, Johnny's Selected Seeds	Check catalog
Patio Varieties	Mini Belle, Chili Pyramid	Thompson & Morgan	Check catalog
	Mohawk, Redskin, Chilli Cheyenne, Chilli Apache	Vegetalis	Check catalog

Continued

TABLE 2.1 (*Continued*)
Plant Varieties and Sources

Plant	Variety	Source	Notes
Tomatoes-Beefsteak (Staking)	Blitz, Match, Quest, Trust, Geronimo	Paramont Seeds	Staking greenhouse variety
Tomatoes-Cherry	Caramba, Favorita, Juanita, Goldita (yellow)	Paramont Seeds	Staking greenhouse variety
Tomatoes-Cocktail	Picolino, Flavorino	Paramont Seeds	Staking greenhouse variety
TOV (tomato-on-vine)	Tricia, Brilliant, Red Delight	Paramont Seeds	Staking greenhouse variety
	Celebrity	Johnny's Selected Seeds, Stokes Seeds	Check catalog
Tomatoes (Bush)	Brandywine-Heirloom	Johnny's Selected Seeds, Stokes Seeds	Check catalog
Tomatoes (Patio)	Tiny Tim, Small Fry, Window Box Roma	Johnny's Selected Seeds, Stokes Seeds	Check catalog
	Totem, Balconi-Red, Balconi-Yellow, Gartenperle	Thompson & Morgan	Check catalog
	Tumbling Tom, Sweet 'n' Neat	Vegetalis	Check catalog

medium such as perlite, vermiculite, or a mixture of these with peat. I have found the best is either vermiculite or perlite. These media have better oxygenation than peat. Coco coir can also be used in place of peat.

Although this method of using multipacks is feasible, it is somewhat inconvenient and messy filling the trays and transplanting. For this reason, most people prefer to use some type of growing cube. You can use Jiffy-7 peat pellets, which are compressed peat contained in a plastic mesh. You soak them in water for 5 to 10 minutes until they swell to about 1 × 1½ inches. You then seed directly into the peat substrate. In some hydroponic systems they may break down and clog your system; I prefer to use other growing cubes such as Oasis Horticubes or rockwool cubes.

Oasis Horticubes come as 1 × 1 × 1½ inch cubes. They are especially good for lettuce, arugula, basil, and herbs. They provide good oxygenation, retain sufficient water so as not to dry out quickly, are sterile, retain their

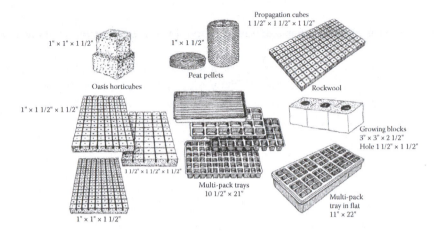

FIGURE 2.1 Seedling propagation cubes, blocks, peat pellets, and trays. (Drawing courtesy of George Barile, Accurate Art, Inc., Holbrook, New York.)

structure fairly well, have a balanced pH, and are easy to handle. They come joined together as 162 cubes per pad that fit into a standard flat. They are easily separated as they are joined only at their bases. I would not use them for tomatoes, peppers, eggplants, or cucumbers as the horticube is very small and not easily transplanted to a larger block.

The best growing cubes and blocks are those of rockwool (Figure 2.2). Rockwool has very good physical properties of drainage, aeration, water

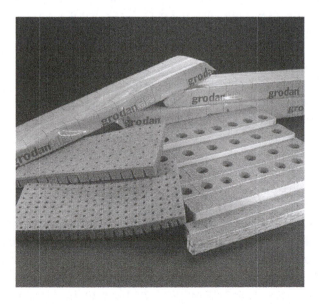

FIGURE 2.2 Rockwool cubes, blocks, and slabs. (Courtesy of Botanicare, Tempe, Arizona.)

retention, structural integrity, and sterility. Chemically rockwool is inert, but slightly alkaline that can be adjusted by an application of water or nutrient solution of optimum pH. They are especially well suited to growing tomatoes, eggplants, cucumbers, and peppers. Lettuce and herbs also do very well in rockwool cubes. They are more expensive than the other methods, but for a hobbyist the extra few cents of cost is more than offset by the convenience and success in germination of your seed.

The propagation cubes come in several sizes. The smaller size cubes of 1 × 1 × 1½ inches high are most suitable for lettuce and herbs. They come in pads of 200 cubes that fit into a standard flat. The larger cubes of 1½ × 1½ × 1½ inches are best for tomatoes, eggplants, peppers, and cucumbers. Pads of 98 cubes fit into a flat. These cubes have small ½-inch holes for placing your seeds. Sow the seeds directly into the holes of the cubes.

Place the cube pads in flats and then soak them thoroughly with water. Soak them prior to sowing the seeds. Be sure that no dry spots are visible after soaking. Use a watering wand on a hose or a hand watering can to soak the cubes. If the seed is fairly new and the viability (percent germination) is 85% to 90%, sow one seed per cube with the exception of herbs (not basil), which you can sow 8 to 10 seeds per cube. Basil should be seeded about 4 to 5 per cube. Always sow only one seed of tomatoes, eggplants, peppers, or cucumbers per cube. You do not need to cover the seed with any substrate; just water the cubes several times a day with a watering can. This will keep the seed moist during it germination. Be careful when you sow the seed that it falls to the bottom of the hole in the cube. For larger seeds like tomatoes, eggplants, peppers, and cucumbers use the back of a pencil to push the seed down should it not fall all the way to the bottom of the hole.

Tomatoes, eggplants, peppers, and cucumbers (vine crops) are transplanted into rockwool blocks after the seedlings grow to a specific age in the cubes. For tomatoes, eggplants and peppers use the 3-inch square by 2½-inch thick blocks. Cucumbers are better transplanted to the 4-inch square by 2½-inch blocks. The blocks have round holes 1½ inches in diameter by 1½ inches deep to allow the 1½ × 1½ × 1½ inch cubes to be placed in them. Blocks, like the cubes, must be soaked thoroughly before transplanting. Tomatoes should be transplanted to the blocks at 14 days (Figure 2.3), eggplants and peppers after 21 days, and cucumbers in about 7 days. Place the blocks in flats in a checkerboard fashion to permit more spacing among the plants; otherwise, they will get "leggy." The cucumbers will have to be planted to the final growing area within a week of transplanting. The tomatoes and eggplants are generally grown until they are 5 to 6 weeks old from seeding before transplanting to the growing system. Peppers may grow a week or so longer than tomatoes before transplanting to the final production system.

FIGURE 2.3 Tomato seedlings 18 days old ready to transplant to rockwool blocks. Note the tomatoes are laid on their sides to permit the plants to grow back upward allowing roots to grow along the base of stems. (Courtesy of CuisinArt Golf Resort & Spa, Anguilla.)

HYDROPONIC CULTURE IN CHOICE OF MEDIUM

The nature of the hydroponic culture system will also determine what medium is the best to use. For example, when growing lettuce in a floating or raft culture system, you need to use a growing cube that will not fall apart upon placing it into the Styrofoam boards.

Although Oasis cubes may be used, I find that they crush easily when pushed into the holes of the boards. This causes material to get into the piping of the system and plug tubes. Rockwool cubes are far superior, as they will not break apart as they are pushed into the holes of the boards. They also stay together well when harvesting the plants. A similar problem may arise when using Oasis cubes in an NFT gutter system, however, less so than with the raft system.

TRANSPLANTING

Transplanting is the next stage in the growing of plants. In some cases you may seed directly into the hydroponic system, but generally it is better to transplant seedlings into the hydroponic system. This allows you to select the best, healthiest plants. Also, if the seed does not have a high germination percentage, you can sow excess seed to obtain sufficient seedling to transplant. If you seed direct you must depend on all of them germinating. In a later chapter I discuss seeding direct into pots of vertical plant towers when using a substrate such as perlite. This is applicable to most herbs. It is especially useful if you want to sow many seeds in one pot, as is the case with the majority of herbs.

With lettuce you can transplant the seedlings in their growing cubes rather than going through a second transplanting, as you would do for tomatoes, cucumbers, and peppers. Whether you are growing lettuce in Oasis or rockwool cubes, be sure that they are ready for transplanting. Lettuce should have at least four to six true leaves before transplanting to the hydroponic system. Depending upon their growth rate, it may take up to 3 weeks from sowing the seed to their transplanting stage. For best results with tomatoes, eggplants, cucumbers, and peppers, transplant the seedlings to rockwool blocks. I shall discuss each separately.

As the tomatoes grow in the cubes, they should be separated after about 10 to 14 days when the first set of true leaves have fully unfolded and a second set begins to develop. Here is an important gardening tip with which most people may be unfamiliar. Separate the cubes of tomato seedlings, space them out in a flat laying the cubes on their sides (Figure 2.4). If you space them correctly you should get about 4 plants across by 9 to 10 rows.

FIGURE 2.4 Transplanting tomatoes to rockwool blocks. Note the plants are placed on their sides in the rockwool blocks. (Courtesy of CuisinArt Golf Resort & Spa, Anguilla.)

Leave about 1 inch between rows and cubes. By laying them on their sides the seedlings will bend up and form adventitious roots on the base of the stem. When transplanting to the rockwool blocks either invert the cubes 180 degrees or place them on their sides (inverted 90 degrees). This will cause the seedling to form roots on the stem that will grow down into the rockwool block as the shoot grows up. The result is a very vigorous, healthy plant with many roots. These extra roots, upon transplanting into the final hydroponic system, will reduce the transplant shock and assist the plant in growing very rapidly. Space the blocks in a checkerboard pattern in the trays. About 10 to 14 days later (about 5 weeks from sowing) they will be ready to transplant to the hydroponic system when they have about three sets of true leaves. If you want to keep them longer before transplanting, space them again to about half the plants (about six plants/tray).

Peppers grow slower than tomatoes, so they will need to be separated as they form the second set of true leaves also, but that will take at least 3 to 4 weeks. Follow the same procedures as with the tomatoes by laying them on their sides at the same spacing in a flat (Figure 2.5). After about 2 more weeks they can be transplanted to the rockwool blocks. You can partially lay them on their sides as you place them in the rockwool blocks. They, similar to tomatoes, will form adventitious roots and should have

FIGURE 2.5 Peppers 25 days old that have been spaced and set on their sides. These plants are ready to transplant to rockwool blocks. (Courtesy of CuisinArt Golf Resort & Spa, Anguilla.)

FIGURE 2.6 Peppers 38 days old (13 days after transplanting to rockwool blocks) ready to transplant to final growing system. (Courtesy of CuisinArt Golf Resort & Spa, Anguilla.)

about three sets of true leaves unfolded when ready to transplant as shown in Figure 2.6.

Eggplants are cared for similarly to peppers, but do not lay them on their sides.

Cucumbers must be cared for somewhat differently than tomatoes and peppers. Cucumbers grow very rapidly. Their leaves expand quickly and to a large size. For this reason, it is important to separate and space them out early. As soon as the first true leaves have unfolded, separate the cubes and space them to about half the number as for tomatoes in a flat (3 × 5 = 15 plants/tray). But do not lay them on their sides, as they are susceptible to such fungus diseases as gummy stem blight that may infect them in a horizontal position. You can in fact skip the step of just separating them in a flat and proceed immediately to transplanting them to the larger 4-inch rockwool blocks. Set them in the blocks upright. Set the blocks in a checkerboard arrangement to get proper spacing (Figure 2.7). You can keep them at this spacing for 7 to 10 days. If you do not transplant them to the final hydroponic system, you will need to space them out again to about half the pants per flat (twice the spacing). Do not hold them in the flats more than about 2 to 3 weeks before their final transplant destination (Figure 2.8). Within a week after transplanting, the cucumbers are growing rapidly (Figure 2.9).

FIGURE 2.7 European cucumber seedlings 14 days old ready to transplant to final growing system. (Courtesy of CuisinArt Golf Resort & Spa, Anguilla.)

SEEDLING TEMPERATURES

I am aware that it may be difficult to maintain optimum temperatures for your seedlings when growing them in your house, as it is not feasible to keep one room where your plants are located at different temperatures from the rest of the house. The only exception may be in your basement. The second difficulty is that you will be growing all of your plants in one area; therefore you cannot have different temperature regimes for different crops. I will, however, give you this information, so that you can use it as a guideline. Tomatoes require from 77°F to 79°F (25°C to 26°C) during germination. As they grow you can lower the day temperature to 73°F (23°C) and the night temperature to 68°F (20°C) for the seedling stage.

Cucumbers like slightly higher temperatures. They take about 2 days to germinate under a day temperature of 79°F (26°C) and night temperature of 70°F (21°C). Upon transplanting to the growing blocks the temperatures may be lowered by about 5°F (3°C) to give a day temperature of 73°F (23°C) and night temperature of 68°F (20°C). That is about the same as for tomatoes.

Germinate peppers between 77°F and 79°F (25°C to 26°C). As the seedlings emerge lower the temperature to 72°F to 74°F (22°C to 23°C). After transplanting use the same temperature regime as for tomatoes.

Lettuce germinates well under temperatures of 59°F to 68°F (15°C to 20°C). Temperatures in excess of 73°F (23°C) may cause seed dormancy. If dormancy is a problem germinate them in a refrigerator at 61°F (16°C).

FIGURE 2.8 Transplant European cucumbers to Bato buckets at 2-week stage. Note the drip lines in the perlite medium and one in the block. Also note the plant clip positioned under the leaves for initial support. (Courtesy of CuisinArt Golf Resort & Spa, Anguilla.)

As soon as the seed breaks place them under lights. The best temperature ranges for herbs is from 65°F to 75°F (18°C to 24°C). Lettuce seedlings can be grown in temperatures a few degrees lower than for tomatoes, but if you are growing your other seedlings in the same area use the tomato temperature range for best results for all seedlings.

LIGHT FOR SEEDLINGS

When growing in your home you need to provide artificial lighting for your plants, including the seedlings. As soon as the seeds germinate, that is they break their seed coats, be careful to give them sufficient light or they will get very "leggy." Lighting will be discussed in more detail in the next chapter. Here all I want to point out is that you need to have at least 5500 lux (510 foot candles) intensity at plant surface for 14 to 16 hours per day. You should purchase a light meter to test that you have sufficient

FIGURE 2.9 Within a week of transplanting the cucumber plants are growing vigorously. Note the tendrils that must be removed daily. (Courtesy of CuisinArt Golf Resort & Spa, Anguilla.)

lighting at the surface of the plants. However, you must not lower the lights too close to the plants or you will add a lot of heat that may force them into leggy growth.

STRAWBERRIES

If you wish to grow strawberries you need to purchase plants; you cannot grow them from seed. Everbearer or Day Neutral varieties will produce under similar conditions to those you will be using for other crops in your hobby unit. Some of the more popular varieties include Seascape, Tribute, and Tri-Star. If you purchase actively growing bare-root plants, then they will produce in about 6 weeks from transplanting. Also tissue-cultured plugs will grow equally as fast. It is best to remove all flowers for the first 3 weeks to allow the plants to get established. Also remove any runners if they form. Otherwise these vegetative structures will take the nutrients away from the parent plant that needs all its energy to produce fruit. Short-day varieties can be grown over the fall-winter if you keep the day length under 12 hours. Short-day varieties are normally sold in the fall. Some short-day varieties include Camarosa, Chandler, Oso Grande, Sweet Charlie, and Douglas.

Also consider if you are growing strawberries in a backyard garden; you could propagate some runners or lift some plants from the garden in the fall and plant them to your hydroponic unit. Store them in the refrigerator for 3 to 4 weeks to induce dormancy. Stressing the plants in the garden will also cause them to go dormant and thereafter be able to resume growth in your hydroponic unit. Stress them by reducing the watering and fertilizing. Transplant them to your hydroponic unit by mid- to late September.

3 Cultural Practices and Environment

PLANT SPACING

Plant spacing is determined by the size of the mature plant. You must consider the actual floor area as the basis for your plant spacing. For example, if you have a small hydroponic unit of dimensions 2 feet by 4 feet that gives you a total area of 8 square feet. Tomatoes require 3.5 to 4 square feet per plant, but that is floor area. You may think that you can only locate two tomato plants in the hydroponic unit, but not so. The roots of most plants can be contained in fairly small areas as long as they receive adequate oxygen, water, and nutrients. In reality you could grow up to eight tomato plants in those 8 square feet of the hydroponic system. But you must train the plants outward in a V-cordon method so that at the top of the support cables the total floor area occupied by those eight plants under your lights would be equivalent to 28 to 32 square feet. That would give the plants adequate spacing so that light could enter the canopy and you would have sufficient access for training the plants.

Peppers and eggplants require similar spacing although their training is somewhat different as will be discussed shortly. You can plant some peppers and tomatoes together in the same hydroponic system. Perhaps the best would be peppers on one end and the tomatoes on the other; do not mix the plants within the row as that may make training more difficult.

Cucumbers (European) require more area, generally 10 square feet per plant. Their leaves are very large, including the lower ones, so you could not grow as many plants in the system. The maximum would be about four, which is half the number of tomatoes, peppers, or eggplants. Four plants would need to be trained so that they occupy 40 square feet of floor area.

Lettuce needs to be spaced according to the actual area of the hydroponic system. Lettuce should be spaced either 6 × 6, 7 × 7, or 6 × 8 inches (Figure 3.1), depending upon variety. Bibb or buttercrunch varieties can be space closer than looseleaf or oakleaf types.

Herbs can be spaced fairly tight. Depending upon their final growth habit they may be spaced 3 × 3 inches or 4 × 4 inches according to the area of the hydroponic unit. Remember to sow up to 10 seeds per cube or direct seed in clumps. Basils and arugula are grown in bunches of 5 to 6 and 10 to 12 plants, respectively. These bunches are spaced at 6 × 6 inches centered

FIGURE 3.1   Bibb lettuce at 6 × 6 inch spacing in Styrofoam board of a raft culture system. (Courtesy of CuisinArt Golf Resort & Spa, Anguilla.)

similar to lettuce. Arugula can also be seeded almost touching each other in rows 6 inches apart.

WATERING

You need to adjust your irrigation cycles as the plants grow. As their leaf area increases and fruit formation is underway the plants will demand more water, so shorten the irrigation cycles. That is, increase the number of irrigation cycles per day. Fully mature tomato plants may use as much as 2 quarts of water a day. But this is also dependent upon temperatures and light intensity. Under indoor growing, irrigate during the mature stage of these vine crops at least once every 2 hours. Of course, this applies to soil-less culture systems that use some substrate such as perlite, vermiculite, sand, or gravel. Nutrient film technique (NFT) and water culture systems have a constant flow 24 hours per day.

Install an automatic float valve system in the reservoir of the unit, so that water may be added to the nutrient tank as the plants use it. This will prevent a possible disaster of running out of solution if you are away for several days. Use a standard float valve assembly similar to your toilet reservoir. You can purchase complete units that are easily installed in any pipe at most irrigation stores.

TEMPERATURES

Optimum temperature ranges differ for each crop. This, however, is not possible for a small indoor unit. It is best to use a temperature regime that is easily achieved in your home as well as being within acceptable ranges for all of your plants. Night and day temperatures should differ by 5°F to 10°F (3°C to 5°C) with night temperature being the lower. For most warm-season crops like tomatoes, eggplants, peppers, and cucumbers, a temperature range of 60°F (16°C) night and 75°F (24°C) day is suitable. For cool-season crops like lettuce night temperature should be about 55°F (13°C) and during the day (light period) 60°F to 65°F (16 to 18°C).

Herbs can withstand a very broad range of temperatures. They will do best, however, at a range similar to that of tomatoes. Under too low temperatures growth will be slowed, whereas under too high temperatures the plants will get "leggy" resulting in soft succulent plants with little fruit formation. In tomatoes the appearance of purple coloration on the undersides of the lower leaves indicates temperatures are too low.

Keep in mind that your lights will also give off heat, raising the temperature near the plants. Locate lights above the crop sufficiently to get the necessary intensity without adding a lot of heat. You can purchase a thermograph to monitor the temperatures on a 24-hour basis. Each chart is good for one week. These instruments are better than a maximum-minimum thermometer having high and low indicators that only tells you what was the maximum or minimum. It does not tell you the fluctuation of the temperature over time, as does the thermograph.

LIGHT

Plants have evolved under natural sunlight. To date no ideal artificial lighting can provide the same quality or quantity of light that the plant would receive under natural sunlight. I wish to outline some types of artificial lights that are available for use indoors so that you may better understand what may be the best for your conditions and the plants you wish to grow. As mentioned earlier you want a minimum intensity at leaf surface in the upper portion of the plants of 5500 lux (510-foot candles) for a period of 14 to 16 hours per day.

TYPES OF LIGHTS

In the past, fluorescent lighting was the most common form of plant supplementary lighting. Some companies promoted Gro Lux lamps specifically for plants. I usually found that cool-white high-intensity fluorescents were the best and cheaper than the plant grow lights. Today, however, lighting for plant growth has become more sophisticated. Most lights sold today are of the high-intensity discharge (HID) type. There are two types of lights: high-pressure sodium (HPS) and metal halide (MH). Usually a

combination of both gives the best results. Also important in your choice
of lights are the reflectors.

HPS bulbs are available in 150, 250, 400, 600, and 1000 watts. MH
bulbs come in 175, 250, 400, 1000, and 1500 watts. The HPS lights provide
more energy in the red part of the spectrum, which promotes blooming
and fruiting, whereas the MH lights are more intense in the blue part of
the spectrum causing rapid growth. Commercial greenhouse growers will
combine MH and HPS lights for seedling growth to be sure that the entire
visible light spectrum is available to the plants. This is not practical for
indoor growing. Hobbyists have found that MH light has a wider spectrum
and therefore is more useful for indoors where no natural sunlight is present. HPS lights are better to supplement sunlight to extend the day length
or to increase intensity during cloudy periods. The lamps cost from $20
to $100.

Some manufacturers now make a "conversion lamp." This permits conversion from HPS to MH. The HPS bulbs are made to operate with MH
ballasts. You can interchange the bulbs using the same ballast and fixture.
The fixtures cost about $200.

Over the past 5 years or so the compact fluorescent lights have been
emphasized to save electrical consumption. Now many of these lights are
sold specifically for the growing of plants. For example, Sunlight Supply,
Inc., markets these compact fluorescent lamps in red, blue, and dual spectrum. These lamps are available in 125-watt, 200-watt, 250-watt, and 300-
watt sizes. The red lamps are recommended for flower production and the
blue for vegetative growing, whereas the mixed spectrum is for plants from
start to finish. These lights, being 110 volt, fit into standard mogul base
sockets so do not require any ballast as do HPS and MH lights. These
lights are placed in highly reflective textured aluminum reflectors to give
maximum reflection and diffusion of the light (Figure 3.2). These bulbs

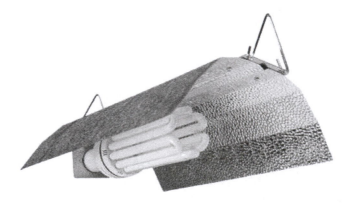

FIGURE 3.2 Compact fluorescent light with reflective textured aluminum
reflector. (Courtesy of Sunlight Supply, Inc., Vancouver, Washington.)

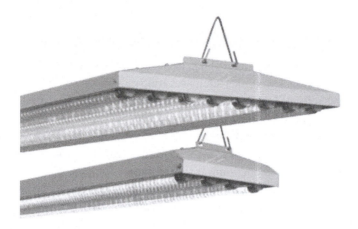

FIGURE 3.3 LED light fixtures with a mixture of four red and four blue tube lamps. (Courtesy of Sunlight Supply, Inc., Vancouver, Washington.)

cost from $50 to $120 depending upon their wattage. The reflectors cost from $40 to $80.

Today, new LED lights are being promoted for their low consumption of power. There are a number of types where the blue and red can be mixed at varying percentages in a reflector unit to provide the best light for specific crops. LED lights for growing plants are available as LED tubes or arrays in various types of fixtures. LED tubes, as manufactured by Sunlight Supply, Inc., are 4 feet in length and fit into 4-foot fixtures. Two fixtures are available, one to contain four LED lamps and the other for eight lamps. They are flexible to mix and match different colored lamps to give the desired spectrum for growing (Figure 3.3). Fixtures range in price from $115 to $190 for the four-tube and eight-tube ones, respectively.

LED arrays, as stated by Apache Tech, Inc., were designed as an alternative to HID lights in greenhouses. They claim that these LEDs last for 50,000 hours (about 6 years). They also state that these light arrays use approximately 80% less power than HID lamps and produce the same intensity. The light arrays are available in white, red, and blue, or as a combination of white and blue, white and red, and red and blue. Color combinations are in the ratios of 1:3, 3:1, and 1:1. There are 120 one-watt bulbs in an array that produces an equivalent of a 1000-watt HID lamp. They work on 120 or 240 volts. The arrays cost from $1000 to $1200 each.

Phillips is also doing research on applying LED lighting to horticulture. Phillips does have some lights that are presently available for use with crops and developed an LED "interlighting" system that provides light between the plants without generating any heat. These lights are being used by some commercial growers to achieve greater control over plants' growth.

AMOUNT OF LIGHT

Most vegetable crops, including herbs, need 50 to 70 watts of light per square foot of growing area. This is a general guideline. I have always found that fruiting vegetables such as tomatoes, cucumbers, eggplants, and peppers like a lot of light, so approach the higher figure for them. To calculate the correct wattage of light needed for a specific area, simply multiply the desired wattage of light by the area in square feet. Using our earlier example from the "Plant Spacing" section, we had 8 tomato plants that would eventually cover 32 square feet of floor space (8 plants × 4 square feet/plant). We need at least 32 × 50 watts = 1600 watts. So you should use a 1500-watt MH light. This 1500-watt MH will provide 47 watts per square foot of growing area. To determine the number of units needed for a specific wattage of bulb, divide the total growing area by the lamp wattage times the desired watts per square foot. For example, if you want to use 600-watt lamps for the 32 square feet of growing area, the number of units required is (32/600) × 50 = 3 (rounded up).

Lights need to be supported by chains. I have found that using jack chains to support them allows you to change their position as the plants grow. Start the plants immediately with the light. Keep the light about 3 feet above to get 510-foot candles at plant surface. Then lower the light about a foot a week until it is about 1 foot above the plants as they grow upward. Then keep moving the light up several times a week in increments of 6 inches as the plants grow up 6 inches. When the light is close to the plants you must take care not to burn them. Lights give off heat, so you must dissipate the heat by using a fan that will move the air across the leaf surface. Remember, as I discussed earlier, you want to keep your day temperatures about 75°F (24°C). You can make more efficient use of the light in the plant canopy by use of a reflector. Most of us just naturally take that for granted. We do not want to have light going up above the plants, just for it to focus on the plants themselves.

REFLECTORS AND WALL COVERING

Reflectors come in three basic forms: parabolic, horizontal, and conical. Claims are made that parabolic reflectors give up to 18% more light than the conical ones. They are supposed to focus more of the light on the plants by directing the light below the horizontal plane. This also reduces glare to your eyes. Conical reflectors give more side light so they are more useful if you want some light to reach out from the plant canopy. Sunlight Supply has parabolic reflectors in 32-inch and 48-inch diameters, and a cone reflector in 42 inches (Figure 3.4). Some reflectors come with a fan to cool

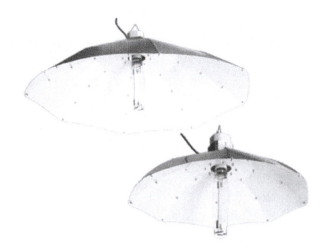

FIGURE 3.4 Parabolic and cone reflectors for HID lights. (Courtesy of Sunlight Supply, Inc., Vancouver, Washington.)

FIGURE 3.5 Square reflector, 8-inch air cooled for HID lamps. (Courtesy of Sunlight Supply, Inc., Vancouver, Washington.)

the bulb and a glass shield to assist in moving hot air away from the lights (Figure 3.5). Square-shaped reflectors are best for square growing areas. For example, Sunlight Supply's 8-inch Blockbuster Reflector (Figure 3.5) covers a growing area of 4 × 4 feet. Sunlight Supply also makes a 4-foot diameter Vertizontal reflector that uses all sizes of MH and HPS lamps (Figure 3.6). It is said to provide a wide area of coverage from low mounting heights. Horizontal reflectors are recommended for HPS systems and where light movers are used. Some of the reflectors accept MH or HPS lamps ranging from 250 to 400 watts. Prices range from $120 to over $200.

Lighting is more efficient when using a reflective Mylar covering on the walls surrounding your hydroponic unit to reflect light back into the plants from the sides. Mylar will reflect up to 98% of the light striking it. It is available in rolls 1 or 2 mil thickness.

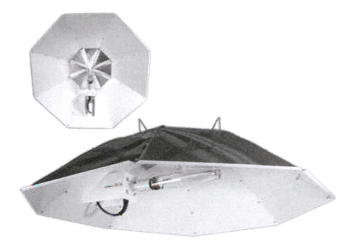

FIGURE 3.6   Vertizontal reflector, 4-foot diameter for HID lamps. (Courtesy of Sunlight Supply, Inc., Vancouver, Washington.)

BALLASTS

To reduce heat buildup near the location of the lights, install the ballasts some distance away from your growing area. Keep them up off the floor in an area free of splashing water and where they will not fall. Connect the ballast into a timer so that you can set the hours of lighting to come on automatically. It takes up to 30 seconds for the bulb to ignite and up to 5 minutes for it to become fully bright. During the warm-up period the light may flicker.

When a MH light is turned off it requires 15 to 20 minutes to cool down. Do not restart it during this cooling period, as that will reduce the longevity of the bulb. Metal halide lights should be replaced every year. HPS lamps need only several minutes to cool. They should be replaced every few years. Prices of ballasts are from $200 to almost $400 depending upon the power and whether they are single or dual fixture (Figure 3.7). Variable watt ballasts that operate 400-, 600-, and 1000-watt bulbs are available at the higher price level. The 1500-watt ballasts usually operate on 240 volts and cost close to $500. Ballasts are not required for compact fluorescent or LED lights. These lights also give instant illumination on startup. The compact fluorescent and LED lights have the great advantage of lasting for many years compared to the MH or HPS lamps.

LIGHT MOVERS

Light can be more evenly distributed over the plant canopy with some form of moving the lights. Light movers will also help to prevent burning

FIGURE 3.7 Multiwattage ballasts for 400-, 600-, or 1000-watt lamps. (Courtesy of Sunlight Supply, Inc., Vancouver, Washington.)

leaves and plants tending to grow toward the lights. Fewer lights can be used to cover the same distance while still keeping the correct intensity by positioning the light closer to the crop canopy without causing burn. Light movers include a drive motor and a rail on which the light travels, or in the case of circular movers the light will rotate around the fixed position of the arms. A linear mover will move the light back and forth on a track. Tracks may be 2 to 6 feet long. The lights travel about 2 feet per minute, but some are available in variable speeds from 2 to 4 feet per minute. They usually have a 0- to 60-second adjustable time delay for pausing the lamp at the ends of its travel to promote even plant growth from end to end. Prices vary from $250 to $300 per unit and to more than $500 for more commercial units that are capable of moving two or three lights side by side as manufactured by Gualala Robotics with its LightRail 5.0 system.

Linear movers are best suited for long, narrow growing areas, whereas circular movers are better for more square growing areas. Revolving 360-degree units may support from one to five lights (Figure 3.8). A three-arm mover with three lights can cover an area of 10 × 10 feet. They revolve at 16 revolutions per hour. These units cost from $350 to $450.

Do not be fooled by claims made that light movers allow you to plant more plants at a higher density than we already discussed. The light movers help to distribute the light more efficiently over the crop, but they do not increase the energy given to the plants; that is, fixed by the wattage and number of lights. However, they can increase production by reducing mutual shading among leaves of the plants, which may result in more even growth.

FIGURE 3.8 Revolving two-unit circular light mover/rotator. (Courtesy of Sunlight Supply, Inc., Vancouver, Washington.)

CARBON DIOXIDE ENRICHMENT

All commercial growers use carbon dioxide enrichment to increase productivity by up to 20%. Although such a system is not critical for an indoor grower, it will increase yields. The ambient level of carbon dioxide is about 300 ppm. In your house it probably is lower than that. Research has found that a level between 800 and 1200 ppm is optimum for most plants. There are several types of carbon dioxide enrichment systems available for hobbyists. A system of bottled gas and small delivery tubes to the plants is feasible but somewhat cumbersome, as the gas tanks have to be refilled. The tanks may be very heavy and awkward to move. A better system is one that generates carbon dioxide by combustion of natural gas (Figures 3.9 to 3.11). These small generators are priced from $400 to $500 depending

FIGURE 3.9 Natural gas or LP-fired carbon dioxide generator. This unit will produce 24 cubic feet of CO_2 per hour at 17,500 BTU consumption of gas. (Courtesy of Green Air Products, Inc., Gresham, Oregon.)

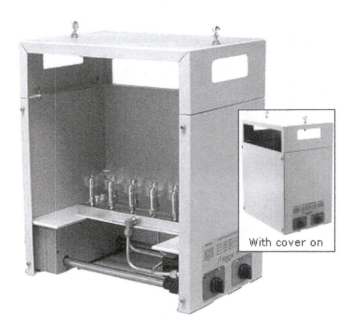

With cover on

FIGURE 3.10 Variable output CO_2 generator with output from 6 to 27 cubic feet CO_2 per hour. (Courtesy of Green Air Products, Inc., Gresham, Oregon.)

upon their capacity. The size of the unit depends upon the volume of the growing space that must be enriched. Variable carbon dioxide (CO_2) generators are also available that can produce from 3 to 6 and 6 to 27 cubic feet per hour of carbon dioxide (Figure 3.10). These range in price from $450 to $600. They are hooked up to your natural gas pipe. They have an electronic ignition control module or safety pilot light shutoff just as your furnace has. Other units produce 13 and 26 cubic feet of CO_2 per hour with capacity to supply grow areas of 15 × 15 feet and 30 × 30 feet, respectively. These cost from $350 to $500. Green Air Products has CO_2 generators (Figure 3.9 and Figure 3.11) that produce 12, 24, or 48 cubic feet per hour that are relatively light in weight at 11 pounds. Their price range is from $400 to $500. All CO_2 generators must operate with a timer only during the light period. Such generators produce as much heat as a 250-watt halide lamp, so you need to move the air with a fan or vent the room to reduce excessive heat. Green Air Products has an optional cooling tube that slides into the units for cooling their generators. Probably during the light period you would be able to heat the plants from the excess heat of the lights and carbon dioxide generator. During the summer months you would have to vent the heated air out of the room to keep temperatures within the desired range.

A more environmentally friendly unit is the CO_2 Boost Bucket system that uses organic compost to naturally generate carbon dioxide

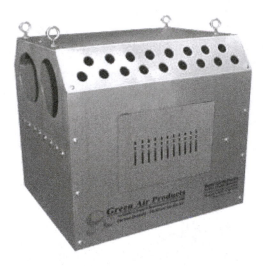

FIGURE 3.11 Carbon dioxide generator capable of 48 cubic feet of CO_2 per hour at 35,000 BTU. (Courtesy of Green Air Products, Inc., Gresham, Oregon.)

(Figure 3.12). One bucket will enrich the atmosphere of a room $10 \times 10 \times 10$ feet (1000 cubic feet) at levels of 1200 to 1500 ppm for a period from 60 to 90 days. After the CO_2 generating organic contents decompose over time and release of CO_2 is reduced, the contents can be used as compost and a new refill is purchased. The kit costs $120 and the refill $100.

Controllers, such as the Green Air Monitor/Controller, enables the user to adjust CO_2 levels within 10 ppm within a range of 0 to 5000 ppm (Figure 3.13). The monitor/controller will activate your enrichment system whenever the level falls below a preset level and turn it off should that level be exceeded. It also comes with a switchable photo sensor to turn off CO_2 production during darkness periods. This unit costs almost $700. There are also CO_2 portable monitors (Figure 3.14) with a measurement range of 0 to 10,000 ppm that costs $700. Very inexpensive test kits that measure from 300 to 5000 ppm cost about $20.

We have been discussing what environmental factors are required to produce healthy plants. Now assuming that you have all of these components for your plants under control, you now need to learn to care for the plants themselves.

FIGURE 3.12   Co2Boost Bucket kit. (Courtesy of Co2Boost, LLC, Landenberg, Pennsylvania.)

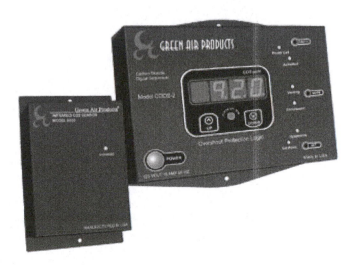

FIGURE 3.13   Carbon dioxide digital sequencer monitor and controller to adjust CO_2 levels within 10 ppm increments from 0 to 5000 ppm. (Courtesy of Green Air Products, Inc., Gresham, Oregon.)

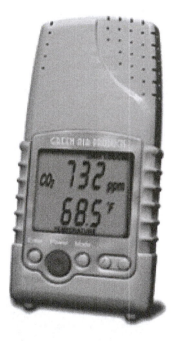

FIGURE 3.14   Portable CO_2 monitor with a range of 0 to 10,000 ppm. (Courtesy of Green Air Products, Inc., Gresham, Oregon.)

4 Caring for Your Plants

Methods for starting your seedlings and transplanting them were described in Chapter 2 "Starting Your Plants." In this chapter, we will look at how to train the plants and some accessories that will facilitate this process.

Tomatoes, eggplants, peppers, and cucumbers must be trained vertically to best utilize the growing area of your room. I use plastic string to support the plants to either a series of hooks or a support wire above. If you are growing the plants in a spare bedroom, you probably do not want to suspend wires with hooks along the ceiling. I believe it is more presentable to use decorative hooks (available at garden centers) that are used for hanging plant baskets from your ceiling. Use one hook for each plant. After all, how many plants are you going to grow? Probably no more than 8 to 10 vine crops. One European cucumber plant that produces two to three fruit per week has always been more than enough for my salads. Several pepper plants and five to six tomatoes should be sufficient for all the peppers and tomatoes of an average family. Remember to grow at least one cherry tomato. I have never tasted any better tomato than the cherries!

STRINGING

Tomatoes, eggplants, and peppers will have to be lowered as they approach the ceiling. Indoors you will probably carry your plants for about 6 to 7 months from seeding. That is, two crops per year. We will talk about that later when discussing cropping schedules. At this point, however, it is important to decide on how long you wish to carry the crops as that will determine how much extra string you should allow for the lowering process. Generally, I have found that tomatoes over a period of 6 to 7 months will grow to a length of about 11 to 12 feet. So, if you have an 8-foot ceiling, allowing 2 feet for your lights, that gives you about 5 to 6 feet of useable height for the plants. Use "tomahooks," which are special wire hooks for attaching to your support cable or hook above (Figure 4.1). Wind 12 to 14 feet of string on the hooks to give you from 10 to 12 feet of string that will support your plants (2 feet is needed from the ceiling to the tops of the plants at their full height). This extra string will permit your lowering of the plants weekly as they reach within 2 feet of the ceiling.

FIGURE 4.1   Plant tomahooks attached to overhead support cable. (Courtesy of CuisinArt Golf Resort & Spa, Anguilla.)

There are two kinds of eggplants and peppers: the standard bush varieties normally grown in the field and the staking greenhouse varieties. Your choice depends upon space and how long you wish to grow any one crop. I have found that the greenhouse varieties are more productive and are more controllable in their training. The bush varieties in a hydroponic system still grow about 3 feet tall, so they need some support with strings as do the tomatoes but without the use of the tomahooks. The greenhouse ones will grow all the way up to your ceiling, similar to tomatoes, but they grow at a slower rate. When cropping them over a 6- to 7-month period, they will reach about 7 to 8 feet in height. For this reason, it is better to use tomahooks with several feet of extra string. The greenhouse pepper and eggplant varieties also need to be trained to two stems, as I shall explain later, so you need two strings per plant. Space the strings in a V-cordon fashion, training each stem in opposite directions.

Cucumbers are trained somewhat differently. Support them with a string tied to an overhead hook but without the tomahooks.

A more convenient string support hook is a Reelenz reel of string with a hook all in one. They come prewound with string making them easier to work with as you simply unreel the string by pulling on it to lower your plants. There is no need to lift the plants and unwind some string from tomahooks to lower the plants.

Begin stringing your plants soon after transplanting to prevent them from falling over. Attach the strings to the stems of the plants using plastic vine clips. As the plants grow, fasten more plant clips about 1 foot apart going up the plant stem.

TRAINING

Tomatoes are trained to a single stem. Support the tomato plant with a vine clip that attaches to the string with a hinge. Locate vine clips about every foot up the plant stem. Be sure to place the clips directly under a strong leaf so that they will not slip down, as shown in Figure 4.2. You can also wrap the string around the stem occasionally to prevent it from sliding down. Always wrap the string in the same direction so you do not forget from one time to the next, which way to go around the stem.

Tomatoes must be trained to a single stem, otherwise they become very leafy ("vegetative") and will not produce a lot of fruit. To do that you must remove all of the side shoots ("suckers") that form between each leaf and the stem (Figure 4.3). Do this at an early stage when the suckers are about 1-inch long. Delaying this pruning will permit a lot of nutrients flowing

FIGURE 4.2   Place stem clamps under a strong leaf.

FIGURE 4.3 Side shoot of tomato plant must be removed early when about 1-inch long.

into this vegetative growth and as they get larger more shock to the plant and difficulty in removing suckers occurs. Anybody who has worked with tomatoes in their garden can tell you that they are very hard on your hands by staining and even causing your skin to crack if you do not wash well after handling them. I always use plastic latex surgeon's gloves when working on tomatoes. Unless absolutely necessary avoid cutting the suckers with pruning shears or scissors as these tools will spread diseases more than your hands.

As the tomatoes begin to form fruit, especially for the first two to three clusters, the fruit load may be very heavy and cause the trusses to break or kink. This will reduce your production as the nutrient flow is interrupted. To avoid this problem use plastic truss hooks to support the fruit truss. There are two types, a C truss hook that has two large C hooks on one end that is attached to the plant stem and a smaller hook on the other end that is attached to the tomato truss (cluster of fruit). The other truss type is a J truss hook that has several barbed hooks on one end that attaches to the support string and a hook on the other end that goes on the tomato truss

FIGURE 4.4   J truss hook that attaches to support string and tomato truss to prevent the fruit truss from breaking.

(Figure 4.4). Usually it is best to have both kinds, as depending upon the length of the truss and tightness of your support string, one may be easier to attach than the other. There is also another type of truss support that is put directly on the truss stem at an early stage to prevent it from kinking (Figure 4.5).

Greenhouse varieties of eggplants and peppers are trained to two stems. As the plant bifurcates at its first flower (crown flower), train these two stems as the principal ones (Figure 4.6). Peppers and eggplants continue to bifurcate at every flower as the stems grow. You must encourage fruit growth by pruning each additional stem to two leaves and flowers. Initially, remove the flowers at the first and second stem layers (bifurcations) to enable the plant to become vigorous. As with your tomatoes remove the suckers often to prevent the plants from getting very vegetative. Support the two stems of the peppers or eggplants with strings and plant clips in a V-cordon way so that the stems are going in opposite directions to permit light to enter the canopy more readily.

European cucumbers are trained somewhat differently than tomatoes, eggplants, and peppers. Cucumbers are supported in a V-cordon

FIGURE 4.5   A plastic truss support that attaches directly to the fruit cluster.

configuration whereby plants are alternately supported to one side and then the other as shown in Figure 4.7. This eliminates leaves overlapping and permits more light to penetrate the plant canopy. They are supported by strings that are directly attached to your ceiling hooks; no extra string is needed. Place one plant clip under the first set of true leaves, but do not pull the sting very taunt when attaching the clip to allow some give as you continue winding the string between each set of leaves as the plant grows upward. Wind the string always in the same direction. Do this every day, as the plants in their growth to the support wire will be about 6 to 8 inches per day. Be careful not to bend the growing tip of the plant too sharply as they can easily break. I usually pull down on the string with one hand while wrapping the plant around the string with the other. If you wind the string in a "clockwise" direction you will remember to do it consistently. In commercial greenhouses you need to be sure that everybody does it the same or one person will wind them up one day and a different person the next day will unwind them unknowingly and the plants will later fall and possibly break.

When the plants reach the top of the support string you need to pinch the growing point. Allow at least two suckers nearest the top of the plant

FIGURE 4.6   Pepper plants are trained with two stems.

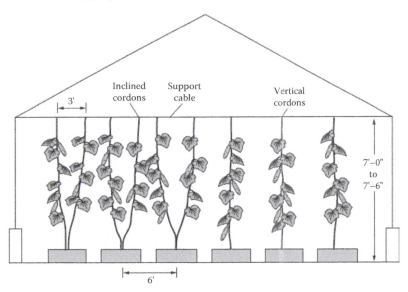

FIGURE 4.7   V-cordon system of training European cucumbers. (Drawing courtesy of George Barile, Accurate Art, Inc., Holbrook, New York.)

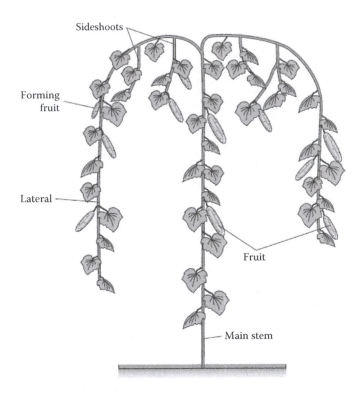

Sideshoots

Forming
fruit

Lateral

Fruit

Main stem

FIGURE 4.8 Renewal umbrella system of training European cucumbers. (Drawing courtesy of George Barile, Accurate Art, Inc., Holbrook, New York.)

to continue growing. This is the renewal umbrella system of training (Figure 4.8). Remove all tendrils as shown in Figure 4.9 (those monstrous stringing growths that wrap around everything) to prevent them from deforming fruits and leaves. Remove all the suckers that form between the leaves and stem on the main stem until the last few near the supporting cable as mentioned earlier (Figure 4.10). Sometimes a second set may form later on the main stem, so be vigilant to prune all of them as soon as they are about 1-inch long. Pull them off by hand. The first seven to eight fruit should also be removed so that the plant can become vigorous enabling it to yield heavily later. Remove the small fruits when they are about 1-inch long (Figure 4.11). Use these as a special delicacy by deep-frying them in batter for a few minutes as you would do with zucchini flowers. They are delicious with the sweet nectar of the flower. The plant will normally only be able to bear four to five fruit at a time. If the fruit set is more, small fruits above will begin to abort until the mature ones are harvested.

Train the top two laterals (side shoots) over the support wire with several plant clips placed about one foot on either side of the main stem and let the laterals grow back down (Figure 4.12). Keep removing tendrils. Usually

FIGURE 4.9 Cucumber tendrils must be removed daily.

the suckers on the laterals will not grow when fruit is forming. Pinch the growing tip of the laterals as they reach two-thirds of the way down or after 8 to 10 leaves. Once the fruit on the first laterals has matured, cut back the laterals to the next set of suckers near the top of the plant and in this way the growth repeats itself with the next set of laterals. With this method of training you can carry the plants up to 10 months. However, in my experience with indoor growing, it is better to start new plants after about 3 months.

POLLINATION

Cucumbers are all female without any male flowers. For that reason, they do not form seeds. We, in fact, do not want any pollination of the female flowers or seed would form. Tomatoes, eggplants, and peppers need to be pollinated.

In commercial greenhouses we employ bumblebees to pollinate. I do not think that you would want these critters in your house; besides, a small hive has far too many bees for your indoor garden, unless you had a large backyard greenhouse. Pollinate by hand. The simplest method is to use an

FIGURE 4.10   Cucumber suckers (side shoots) must be removed.

FIGURE 4.11 Small cucumber fruit about 1-inch long need to be removed from the plant up to 7 leaf axils to promote initial vegetative growth.

FIGURE 4.12 Two laterals trained over the support cable in renewal umbrella training. (Courtesy of CuisinArt Golf Resort & Spa, Anguilla.)

FIGURE 4.13   Use a Petal Tickler or electric toothbrush to pollinate the tomato flowers. (Courtesy of CuisinArt Golf Resort & Spa, Anguilla.)

electric toothbrush (Figure 4.13). During the late morning or early afternoon is the best time to pollinate. At that time the relative humidity will be lower and the pollen will flow freely. For 3 to 4 seconds hold the vibrating toothbrush on the truss (cluster) of flowers and watch that the entire cluster moves at high speed. This vibration will release the pollen, which can be seen as a fine yellow powder. Pollination is done when the flowers are receptive. This is evident from the flower petals curling back (Figure 4.14).

Pollinate at least every other day when the relative humidity is close to 70%. If the relative humidity is very high, the pollen will be damp and sticky, so will not flow readily. Under very low relative humidity of less than 60% the pollen may dry. Close to midday is generally the best time to pollinate.

You will be able to determine whether your pollination is successful within about a week, as small beadlike fruit will start to form (Figure 4.15). This is termed fruit set. It is very important to get your plants into a generative state once they start blooming, that is, they will shift their nutrients

FIGURE 4.14 Receptive flowers of tomatoes ready to pollinate.

FIGURE 4.15 Fruit set on tomatoes.

into more fruit production than just leaf and stem growth (vegetative state). To assist this shift from vegetative to generative state, pollinate every day as the first three flower clusters form. Once these first three clusters have set fruit the plant will become more generative in its growth phase.

Peppers and eggplants should also be pollinated but with less vigor to avoid breaking the flowers. Installing a horizontal airflow fan in the growing room will also assist in pollination. As we discussed earlier, such a fan is needed to move any heat generated by our lights away from the plants to prevent burning and high temperatures within the plant canopy.

PLANTING SCHEDULES

In indoor hydroponics there is little affect of the seasons on your growing. Schedule your cropping around the convenience of your needs for fresh products. We want product most during the winter months when high-quality vegetables are not available in our local supermarket.

In my experience I have found that it is best to grow lettuce and herbs in a separate system from cucumbers, tomatoes, and peppers. This is discussed in more detail in a later chapter on specific hydroponic units. The main reason for growing the lettuce and herbs separately is that they are very low in form and would not grow well under vine crops when your lights are situated near the ceiling as the vine crops mature. Lettuce also yields greater under a very different nutrient formulation from the vine crops. Finally, the lettuce is on a very short cropping cycle of 30 to 40 days after transplanting, whereas the vine crops continue growing for up to 6 months or longer.

Lettuce can be sown every few days to provide about three transplantings per week. This will give you lettuce every day to harvest. The number of plants to grow depends upon your personal demand for fresh salads.

Herbs, with the exception of basil, grow fairly slowly initially. They take about 3 to 4 months to get well established. Once they are growing vigorously, you can harvest them every day. They will continue to grow for a full year. Basil needs about 6 weeks to get well established. By keeping the basil well pruned from the beginning the plants will last up to 3 to 4 months. As soon as the basil reaches 3 to 4 inches tall pinch the growing tip or cut it with a scissors after the second to third node. This early first cut will make the plants branch out more than if cut later and it will prevent the plants from getting woody. Thereafter, keep cutting the tops of each shoot back by 3 inches to the location of the next or lower set of small shoots forking between the stem and leaves. Pruning in this fashion will give you a plant that has many branches and will as a result have less woody growth (Figure 4.16). That is the secret I have found with growing basil. In fact, I have grown basil for up to 6 months or longer by consistently pruning (harvesting) it. If it begins to flower, it is getting old or under stress. Pinch all flower buds very early to keep it more juvenile.

Normally, sow about four plants per cube to make a bunch or cluster of plants. These can be transplanted to 6 × 6 inch spacing in the growing bed.

Assuming you want tomatoes, peppers, and cucumbers most during the winter from November through March, plan the crop accordingly. Tomatoes need to be started no later than mid-August to begin harvesting fruit by mid- to late November. A 7-month cropping cycle takes them to the end of March. Tomatoes require about 100 days from seeding to first harvest. Begin a second crop by sowing seeds in early March for harvesting to commence in June.

FIGURE 4.16   Pruning basil early results in multistemmed plants.

Greenhouse peppers take about a month longer than tomatoes to mature. It would be best to start them in July and again in March. You can use bush varieties that take about a month less to mature than the greenhouse ones.

European cucumbers require 2 months from seeding to first fruit production. However, if you are going to grow all of these crops together in the same hydroponic unit, start the cucumbers in September so that cucumbers will be ready by November. In this way, the seedlings that are transplanted to your hydroponic system will all start out at the same size and grow up together. This will assist in raising the lights at the correct level above all the plants. It also prevents any older plants from shading out the younger ones.

Overall, then, you will have two crops a year. Start the seedlings in a separate system under their own lights, so as not to interfere with the existing crop as you make the changeover.

CROP CHANGEOVER

About one month prior to the scheduled end of your crops, cut off the growing point of your tomato plants. This will shift nutrients to the forming tomatoes. Continue to remove any suckers that develop near the tops of the plants. There is no need to do this for the cucumbers and peppers.

Several days prior to removing the plants, spray them with a mixture of insecticides and fungicide to control the presence of any insects or diseases. After this period, one day prior to pulling the plants you can stop the flow of nutrients to wilt the plants, reducing their weight and volume to facilitate their removal. Compost the old plants if you have an outdoor garden or place them in plastic garbage bags for disposal. Do not keep the string, but you can use the plant clips again if you soak them for a day in a 10% bleach solution. Wipe the walls and ceiling with the same disinfectant or Virkon. If you are using a Mylar reflective surface on the walls be sure to clean it thoroughly or better yet remove it and replace it with new material.

Sterilize the hydroponic unit with a 10% bleach solution. Run the solution through the system to clean the irrigation lines. Let it sit in the lines for 24 hours. Then rinse the lines with fresh, clean water several times. If you find some salt formation in the irrigation lines, flush them with an acid and water mixture in the ratio of 1 to 50, respectively. Flush the system thoroughly several times, and then let it sit for 24 hours before rinsing with water.

Change the substrate rather than trying to sterilize it. That is much simpler than trying to sterilize it by baking it in the oven of your stove. Besides, I find that rockwool and even perlite break down in structure during the sterilizing process.

CONTROLLING PESTS AND DISEASES

You may think that it is impossible to get insects and diseases in a well-protected area like a room or basement of your house, but think again; nothing keeps these critters out. Certainly, you can reduce such occurrences by keeping the room isolated from the outside movement of air but that will not exclude all pests. Keep a close eye on your plants; learn to identify the most common insect pests and diseases. Hang some Bug-Scan cards near the crops (Figure 4.17). These are yellow sticky cards that attract the insects. They work like the old standard flypaper. Once a week identify and count the number of each insect stuck to the cards. Keep this information in a table in a book so that you can refer to it to determine the changes in numbers of insects developing over time. As soon as you discover some pests, you need to take action to reduce or eliminate them, or they will get out of hand, feeding on your plants leaving little product for yourself.

Some of the most common pests that you have to identify are white-fly, aphids, two-spotted red-spider mite, thrips, leaf miners, fungus gnats, and caterpillars (larvae of butterflies and moths). Whiteflies are by far the most common pest of tomatoes, eggplants, peppers, and cucumbers and even lettuce. To learn to identify them correctly and understand how they develop, refer to books on pests. Some are given in the references in Chapter 8. In addition, there are a number of Web sites listed that have

FIGURE 4.17   Bug-Scan card to monitor insect populations. Note the grid and insects on the card.

photos of these insects. Clear drawings and life cycles are also given in my book *Hydroponic Food Production*.

The most difficult part is to determine what control measures to use on these pests. I hope to simplify that for you here with an outline of some of the safe bioagents that I have used very successfully. Most are available from your local hydroponic shop or from other greenhouse suppliers. Also you may use beneficials. These are insects that are predators or parasites on the insect pests. This combined with the use of safe natural bioagents make up an integrated pest management (IPM) program that is purely biological control.

Briefly I am going to discuss the beneficials that help to control a number of the pests. Whiteflies are parasitized by *Encarsia formosa* and *Eretmocerus eremicus*, parasitic wasps. The two-spotted mite can be controlled with the predators *Phytoseiulus persimilis*, *Metaseiulus occidentalis*, and *Amblyseius californicus*. *Metaseiulus occidentalis* is more active under cooler temperatures and *Amblyseius californicus* prefers warmer temperatures. Aphids are kept in check with various lady beetles and the green lacewing, *Chrysopa carnea*. *Aphidoletes aphidimyza*, a midge larva,

FIGURE 4.18 *Amblyseius swirskii* introduced with a bran substrate on peppers for control of whiteflies using a shaker bottle. (Courtesy of CuisinArt Golf Resort & Spa, Anguilla.)

is also an effective predator. Vertalec, a parasitic fungus, *Verticillium lecanii*, may also be used. Several wasps— *Dacnusa sibirica, Diglyphus isaea,* and *Opius pallipes*—parasitize leaf miners. *Orius tristicolour* and *Hypoaspis miles* feed on thrips. These beneficial insects come in plastic shake-on bottles or paper strips of pupae that you hang from the leaves of your plants (Figure 4.18 and Figure 4.19). You may purchase these beneficial insects at your local hydroponic shop or from greenhouse suppliers. A very useful reference book on recognizing and determining appropriate beneficials is *Knowing and Recognizing*. It is listed in the reference section of Chapter 8.

Other pests may be controlled with various beneficial microorganisms. Caterpillars are very common on lettuce. Control them with Dipel or Xentari, which is a parasitic bacterium, *Bacillus thuringiensis*. Fungus gnats may also be kept in check with *Bacillus* bacterium such as Gnatrol. An insect-parasitizing nematode, *Steinernema carpocapsae*, controls fungus gnats.

Extracts from bacteria and plants assist in controlling numerous pests. Agrimek and New Mectin control mites. Azatin or Neem-X, a plant extract from the neem tree, can be used on whiteflies, armyworms, and mealybugs. Cinnamite, a cinnamon extract, is good on aphids and mites. Pyganic, a pyrethrin, is good for aphids, caterpillars, leafhoppers, and whiteflies. Entrust, a spinosad bacteria extract, controls caterpillars, thrips, and leafminers. Fulfill, a pymetrozine, is effective on aphids and

FIGURE 4.19 *Encarsia* pupae on paper strip is hung on a tomato leaf.

whiteflies. BotaniGard, which is spores of a beneficial fungus, *Beauveria bassiana*, is used for whiteflies, aphids, thrips, mealybugs, and weevils. It is a fungus that enters insects and parasitizes them. Pyrethrums come from the pyrethrum daisy. They are effective contact insecticides against a broad range of pests. Spintor controls thrips. M-Pede, a soapy material from salts of fatty acids, is a contact insecticide effective on aphids, mites, thrips, and whiteflies. Fatty acids are the main component of the fats and oils found in plants and animals. M-Pede is also useful as a sticker for applying other pesticides.

The addition of one tablespoon of brown sugar per gallon of spray will assist as a sticker/attractant for many of these products. It slows runoff of the spray from the leaves and may act as an attractant for some insects such as thrips, which hide in the growing point of the plant and therefore are difficult to contact. Bugitol, an oleoresin of *Capsicum*, such as peppers, is used to control and repel such insects as aphids, armyworms, leafhoppers, leafminers, thrips, and whiteflies. Ultra-Pure Oil is petroleum oil that is recommended for control of aphids, mites, leafminers, leafhoppers, mealybugs, thrips, and whiteflies. There are many new natural insecticides coming out, so keep a watch for them on the Internet and horticultural magazine advertisements.

Many of these bioagents are compatible with beneficials as they are selective in their action.

Bioagents are also used in the control of diseases. Phyton-27 is composed of natural minerals and is effective as a fungicide and bactericide. Cinnamite offers control of powdery mildew, a fungus disease. Elemental sulfur combined with M-Pede as a sticker is very effective against powdery mildew. RootShield and PlantShield are from a beneficial fungus, *Trichoderma harzianum*, which act as a protectant against root diseases of plants. Kocide 2000 is a copper compound that helps to control powdery mildew. AQ 10 is a biofungicide that controls powdery mildew. Oxidate, a form of hydrogen dioxide, is effective in controlling root diseases such as damping off organisms that cause death of seedlings. It, like RootShield, is applied to the substrate as a drench before seeding or transplanting. Actinovate is *Streptomyces lydicus,* is used as a fungicide similar to Oxidate.

I do not want to give you exact rates for usage of these products as they may differ for different makes and situations under which they are applied. Always read the directions and follow them exactly! Do not be alarmed by this long list of pest problems. Normally, you may be troubled with a few, unless you do not observe your plants and just let things get out of control. Wash your hands and clean your shoes before entering your growing room. Placing a bleach mat at the entrance through which you must walk prior to going into the room will help prevent bringing in organisms from outside. Special mats are available. Keep the mat moistened with a 10% bleach solution.

5 Plant Nutrition

ESSENTIAL ELEMENTS

Plants require 13 essential elements for their growth. In addition to these 13 nutrients they utilize carbon, hydrogen, and oxygen, which come from water and the atmosphere. The 13 essential elements are categorized in two groups: (1) those needed in relatively large amounts, termed major or macroelements; and (2) those consumed in relatively small amounts, which are called micro or trace elements. The macroelements include nitrogen (N), phosphorous (P), potassium (K), calcium (Ca), magnesium (Mg), and sulfur (S). The microelements are iron (Fe), manganese (Mn), copper (Cu), boron (B), zinc (Zn), molybdenum (Mo), and chlorine (Cl). Plants cannot live without any one of these elements, hence the term "essential." We as growers must provide all of these 13 nutrients to the plant. In hydroponics they are all added in the nutrient solution.

I often encounter the argument that hydroponic plants are not "organic" because in hydroponics you use fertilizer salts. My reply is that all plants are organic as they only use inorganic ions in their uptake of these elements to grow. The plants manufacture the organic materials of their makeup through photosynthesis. "Organic gardening" is often confused with the fact that in most cases it indicates that only natural bioagents are used to control pests rather than synthetic pesticides. As a result, organic products should really be termed "pesticide free."

SOIL IN COMPARISON TO HYDROPONICS

Soil has both organic and inorganic components. The organic part is the humus consisting of dead plant and animal matter. The inorganic component is the sand, gravel, and rock that must be weathered to release its inorganic elements. The organic material undergoes decomposition by numerous soil organisms and animals. This releases its inorganic elements into the soil water. The released inorganic elements into the soil water give us the soil solution. This soil solution comes in contact with the plant roots where the individual elements in their atomic state are taken up through a process of electrical and chemical potential difference between the inside and outside of the root membrane (Figure 5.1).

In hydroponics we add all of these essential elements from highly pure fertilizer salts. They dissolve in water releasing their individual elements

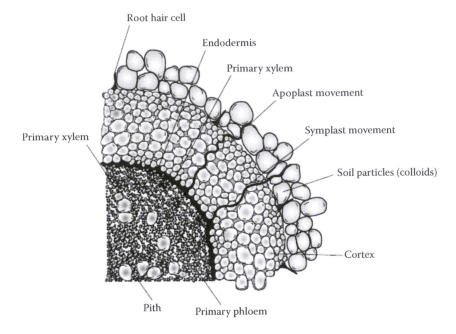

FIGURE 5.1 Cross-section of root with uptake of water and minerals from the nutrient solution into the vascular system. (Drawing courtesy of George Barile, Accurate Art, Inc., Holbrook, New York.)

in an atomic ionic state. This is the nutrient solution that contacts the plant roots providing the elements that are taken up with differential uptake in the same way as for the soil solution. The difference between hydroponic and soil cultivation is that in hydroponics we add exactly the correct amounts in the optimum ratios of each element so that plant growth is not restricted as it might be in soil when some of these essential elements may be at nonoptimum levels. When growing hydroponically, we use bioagents and beneficials as discussed in the previous chapter so our products are pesticide free. This is termed integrated pest management (IPM).

SOURCES OF ESSENTIAL ELEMENTS

In most cases the indoor hydroponic hobbyist will purchase his nutrients prepared from one of the hydroponic outlets. Generally, the nutrients come in two components, an A and B formulation. Two separate components prevent any chemical reaction from taking place in their concentrated form. Usually one part will contain calcium, nitrogen, potassium, and iron, whereas the second one contains the remaining elements. Dissolve them in water separately to prevent any precipitation. The normal ingredients contained in part one (A) include calcium nitrate, potassium nitrate, and iron

FIGURE 5.2 Complete liquid hydroponic nutrient solutions. Grow formulation on the left, Bloom in the middle, and Ripe formulation on the right. (Courtesy of Botanicare, Tempe, Arizona.)

chelate. The second component (B) includes potassium sulfate, monopotassium phosphate, magnesium sulfate, manganese sulfate or chelate, zinc sulfate or chelate, copper sulfate, boric acid, and ammonium molybdate.

Today many single complete formulations as well as ones with A and B parts are available in liquid form. Some are composed of the basic fertilizer salts that we normally use in hydroponics. Many others are of organic base, such as fish fertilizer, bat guano, humic acid, seaweed extract, feather meal, kelp, alfalfa extract, and molasses, and other organic derivatives by themselves as pure organic nutrients or in combination with some fertilizer salts and micronutrients. When selecting a nutrient, look for those specifically formulated for hydroponic use. There are numerous nutrients that are suited specifically for the various plant stages of growth, particularly for the initial vegetative growth stage, followed by bloom and ripening stage formulations to promote flowering and fruit production. For example, Botanicare markets CNS17 Grow Formula 3-2-4, Bloom Formula 2-2-5, and Ripe Formula 1-5-4 (Figure 5.2). General Hydroponics has a FloraNova series of nutrients with a FloraNova Grow (7-4-10) and a FloraNova Bloom (4-8-7) for the different growth stages of the plants. These are a few of the nutrient suppliers that are listed in Chapter 8 in Table 8.1.

Hydroponic shops also sell a wide range of plant supplements and additives (Figure 5.3). Mycorrhizae inoculants are beneficial fungi that live symbiotically on plant roots (Figure 5.4). The fungal filaments (hyphae) extend into the substrate particles and extract, transport, and assist in the plant roots' efficiency of nutrient and moisture uptake. Some of these

FIGURE 5.3   Many plant nutrient formulations and supplements are available at hydroponic shops. (Courtesy of Botanicare, Tempe, Arizona.)

FIGURE 5.4   Soluble Mycorrhizae beneficial fungus. (Courtesy of Plant Success, www.plant-success.com.)

FIGURE 5.5   Silicate additive of potassium silicate to strengthen plant growth. (Courtesy of GroTek, Langley, B.C., Canada.)

mycorrhizal root inoculants include both endo- and ectomycorrhizal fungi. With an expanded root system in the presence of mycorrhizae, enhanced nutrient and water uptake gives plants faster growth, higher yields, and protects them against diseases and pests. Some manufacturers include bio-stimulants such as sea kelp, humic acid, and vitamins C, B, and E with the inoculants to ensure prolific colonization. These products are in liquid, granular, or powder form that dissolves into water or nutrient solution. Silicone is a plant element that is abundant in all soils but can be quite low in many soilless media. Silicate additives containing potassium silicate rectify this deficiency and increase plant resistance against pests and diseases by strengthening plant cell walls resulting in a stronger plant that resists insects, mildews, and fungi (Figure 5.5). It is added to the nutrient solution or applied as a foliar spray.

If you wish to become more involved in the actual nutrition of the plants you may make up your own formulations. However, to do so you need to purchase and store some bags of fertilizers. Do you remember a little about high school chemistry classes that perhaps bored you? Here is your chance to apply it. We will use the chemical symbols for the various salts and units of concentration as parts per million (ppm) or milligrams per liter (mg/l). A general formulation and weights per 10-U.S.-gallon nutrient tank are given in Table 5.1. You will need a triple-beam balance to weigh these salts in gram units.

Table 5.1 includes only the macroelements. The weights are very small for a 10-gallon tank, so you must use caution not to make a mistake in

TABLE 5.1
A General Macronutrient Formulation for 10-U.S.-Gallon Tank

Fertilizer Salt	Weight/10-Gallon Tank (gm)	Elements Supplied and Concentration (ppm)
Calcium nitrate (Ca(NO$_3$)$_2$)	31	Ca: 180 ppm, N: 126 ppm
Potassium nitrate (KNO$_3$)	4	K: 39 ppm, N: 14 ppm
Potassium sulfate (K$_2$SO$_4$)	23	K: 250 ppm, S: 102 ppm
Magnesium sulfate (MgSO$_4$)	19	Mg: 50 ppm, S: 66 ppm
Monopotassium phosphate (KH$_2$PO$_4$)	8.5	P: 50 ppm, K: 63 ppm
Totals		N: 140 ppm, P: 50 ppm, K: 352 ppm, Ca: 180 ppm, Mg: 50 ppm, S: 168 ppm

weighing the compounds. The weight of the microelements will be even smaller, so we need to make a concentrated stock solution with them and then just add some of that prepared solution to your tank. You can purchase a 10-gallon gasoline or water tank to store the micronutrient stock solution. Keep it in the dark away from direct heat to prevent algae growth. Table 5.2 gives you the weights for a micronutrient stock solution of 300 times normal strength in a 10-gallon container.

For a 10-gallon nutrient tank you need to add: $10 \times (1/300) = 0.0333$ U.S. gallons of the micronutrient stock solution. That is in liters: $0.0333 \times 3.785 = 0.1262$ liters or 126 milliliters (ml) (1000 ml = 1 liter). Purchase a 100-ml graduated cylinder to measure this volume of solution. If you cannot get the triple-beam scale and graduated cylinder from your local hydroponic shop, contact a scientific laboratory supply company, such as Fisher Scientific.

TABLE 5.2
A 300 Times Strength Micronutrient Stock Solution in 10-U.S.-Gallon Tank

Compound	Weight/10-Gallon Tank (gm)	Elements Supplied and Concentration (ppm)
Manganese sulfate (MnSO$_4$)	41	Mn: 0.8 ppm
Copper sulfate (CuSO$_4$)	3.2	Cu: 0.07 ppm
Zinc sulfate (ZnSO$_4$)	5.5	Zn: 0.1 ppm
Boric acid (H$_3$BO$_3$)	20.5	B: 0.3 ppm
Ammonium molybdate ((NH$_4$)$_6$Mo$_7$O$_{24}$)	0.6	Mo: 0.03 ppm

We are still missing iron (Fe), which is also a minor element. Using iron chelate (FeDTPA) that has 10% elemental iron, we need 2 grams in a 10-gallon tank to add 5 ppm of iron. You now have your nutrient formulation complete with all 13 essential elements.

WATER ANALYSIS

Before making up the nutrient solution you should have a water analysis of the local water supply to determine the levels of these 13 elements in the raw water. Adjust your formulation for the presence of any of these elements in the water. For example, if the raw water has 10 ppm of calcium, then simply subtract this amount from the original formulation. All of these elements are in ratios, so it is easy to calculate any changes.

Taking our example of 10 ppm of Ca in the raw water, then we want $180 - 10 = 170$ ppm of additional calcium using calcium nitrate. We adjust the formulation as follows: $(170/180) \times 31$ gm = 29 gm of calcium nitrate. At the same time the level of nitrogen provided by the reduced calcium nitrate will fall to $(29/31) \times 126$ ppm = 118 ppm. That is, there is 8 ppm less of nitrogen. That would not influence our optimum level of nitrogen. If there were a larger drop of greater than 10% we would need to add that from another source such as potassium nitrate.

pH OF NUTRIENT SOLUTION

The pH scale measures the relative acidity or alkalinity of a solution or medium. The scale has a range from 0 (extremely acid) to 14 (extremely basic). These sorts of levels you will not encounter. Most plants prefer a slightly acid condition between 5.8 and 6.5. The pH affects the plants' ability to take up its essential elements from the nutrient solution. It also influences the solubility and capacity for the nutrient solution to retain its individual elements in solution.

There are a number of ways to monitor the pH of your nutrient solution. Your choice of testing system depends upon the amount of accuracy and reliability in detection you require. Although sophisticated pH test meters may be accurate within 1/100 of a pH unit, they really are not practical for other than laboratory use. They also need constant calibration and are very delicate to handle. They will cost from $300 to $500 and upward. Simpler, more durable handheld pH meters and "pen" types that have accuracy within 0.1 pH units are available for under $100 to $200 (Figure 5.6 and Figure 5.7). They too, however, must be calibrated frequently and always kept moist.

My preference is to use good pH indicator paper that has a range between 4.0 and 7.0. This paper is accurate to within 0.3 units. My favorite is Merck

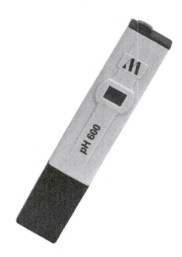

FIGURE 5.6   Pen-type portable pH meter. (Courtesy of Milwaukee Instruments, Rocky Mount, North Carolina.)

FIGURE 5.7   Bluelab pH meter with calibration mixes. (Courtesy of Bluelab Corporation, Ltd., Tauranga, New Zealand.)

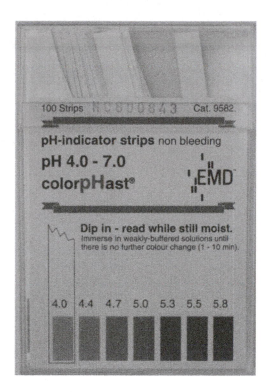

FIGURE 5.8   pH indicator paper.

colorpHast indicator strips as shown in Figure 5.8. These cost about $15 for 100 strips. Paper strips are much easier to handle than liquid dyes that have to be added to a solution sample and then the resultant color of the liquid compared with a color pH chart.

Test the pH of the nutrient solution every day and keep records of it, so you may see any changes taking place. To lower the pH, add acid such as sulfuric acid (H_2SO_4) (battery acid) and to raise the pH, add potassium hydroxide (KOH), sodium hydroxide (NaOH), or baking soda. You must wear gloves and protective goggles when using sulfuric acid, potassium hydroxide, or sodium hydroxide as they burn your skin. These, however, are most effective in quickly shifting the pH. If you buy them in concentrated form, dilute them to about 1 part of acid to 30 parts of water. Remember: add acid to water!

Your local hydroponic shops also sell more dilute solutions to adjust the pH that are safer to handle. These are usually termed "pH UP" or "pH DOWN" solutions (Figure 5.9). These solutions enable the gardener to maintain optimum pH levels of the nutrient solution. Most are in a liquid form, but some are available in powder form.

FIGURE 5.9 pH Up and pH Down solutions for adjusting the pH of nutrient solutions. (Courtesy of Technaflora Plant Products, Ltd., Mission, British Columbia, Canada.)

ELECTRICAL CONDUCTIVITY (EC) OF THE NUTRIENT SOLUTION

Electrical conductivity is a measure of the nutrient solution's concentration through its ability to conduct electricity. Pure water does not conduct electricity, but any water having solutes (elements) added to it has the capacity to conduct electricity. A special meter, an electrical conductivity (EC) meter (Figure 5.10), measures the electricity conducted by the nutrient solution, which is directly related to the level of total dissolved solutes in the solution. The scale is commonly expressed as millimhos (mMhos) or millisiemens (mS).

You cannot directly relate parts per million (ppm) to EC as different ions conduct electricity at different efficiencies; therefore, the ratios of these ions as well as their individual concentrations in the solution will influence the electrical conductance. Electrical conductivity does inform you of the overall concentration of all elements within the solution so it can be used as a guide to tell whether the solution has adequate nutrients. But you could still have an imbalance of individual ions, as the solution may have a high

FIGURE 5.10   An electrical conductivity (EC) meter to measure total dissolved solutes in a nutrient solution.

level of a very highly conductive element such as potassium and a low level of one that conducts less electricity such as nitrate ions. So, remember, the EC gives you only an idea of the overall total dissolved salts within the solution. To determine the exact levels of each element, you would have to submit a sample of the nutrient solution for atomic absorption analysis in a private laboratory. Such an analysis would cost about $45 to $60.

EC meters are essential for anyone growing hydroponically, as they assist in monitoring the nutrient levels of the solution. There are many makes on the market from pen types that cost about $100 to combination pH, EC, total dissolved solutes (TDS), and temperature testers that cost about $200 to the more sophisticated handheld portable meters in the range of $200 (Figure 5.11). These meters are often 3-in-1 meters for testing pH, EC, and TDS (Figure 5.12). Generally, a nutrient solution may have an EC of 1.5 to 3.0 mMhos.

The best is to test the solution as soon as you make it. Record this value, then test and record the EC every day. That will tell you of weakening trends in the nutrient solution as the EC goes down over a period of several weeks. At some point, often within a month, the EC will fall enough to justify changing the entire nutrient solution. The EC meter then indicates at what time you need to change the solution before deficiencies may occur. With experience and a few solution analyses by a laboratory you can relate

FIGURE 5.11   A pen-type hand-held combination pH/EC/TDS/temperature tester. (Courtesy of Hanna Instruments, Woonsocket, Rhode Island.)

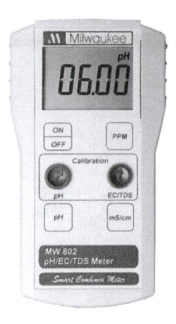

FIGURE 5.12 Milwaukee portable 3-in-1 pH/EC/TDS meter. (Courtesy of Milwaukee Instruments, Rocky Mount, North Carolina.)

the changes in EC with certain elements that consistently get used up more than others. Then you can carry the solution longer by making periodic additions of these elements based on the changes in EC readings.

SYMPTOMS OF NUTRITIONAL AND OTHER PROBLEMS

If your plants are receiving not enough or too much of any element they will show symptoms of yellowing (chlorosis), browning (necrosis), deformation, or stunting of growth. For specific descriptions, functions of elements within the plant, and determination of nutritional disorders, refer to my book *Hydroponic Food Production*. Here I only want to point out some generalities to assist in recognizing nutritional problems with your plants.

Be aware that diseases or insects may also cause symptoms of spots and chlorosis. The first thing to do is to determine whether the symptoms are on the lower or upper part of the plant. The essential elements are grouped as those that are mobile, which can be retranslocated, and those that are immobile, which cannot be retranslocated (moved) to another part of the plant. If they are immobile the first symptoms will appear on the upper part of the plant. Mobile elements express their deficiency on the lower part of the plant as they move to the new growth leaving the lower leaves to suffer. Often fruit formation may be affected at the same time. For example, one of the most common symptoms is blossom-end rot (BER) as shown in

FIGURE 5.13   Blossom-end rot (BER) of tomatoes.

Figure 5.13. This is normally caused by insufficient calcium. It results in tissue breakdown producing a dry, black, leathery appearance of the fruit on the blossom end. The growing points of the plants at the same time will stop expanding and in severe cases will eventually die.

However, environmental factors or watering may also contribute to growth problems. Insufficient watering, too long of periods between irrigation cycles, may also cause BER. If plants wilt they will probably form BER of fruit. So the calcium deficiency in this case is not caused by insufficient calcium in the nutrient solution but by inadequate watering. Another symptom of insufficient or too frequent irrigation cycles is fruit cracking (Figure 5.14). This may occur more frequently under very intense light or excess temperatures. Cool temperatures and high relative humidity, especially at the base of the plant, can cause deformation of fruit or what is termed "catfacing" (Figure 5.15).

These are only a few examples to make you aware that plant growth disorders may be a result of nutrition, pest, disease, environmental conditions, or watering. Be vigilant to recognize symptoms and relate what may be their cause early in the expression of any growth abnormalities. Then you

FIGURE 5.14   Fruit cracking of tomatoes.

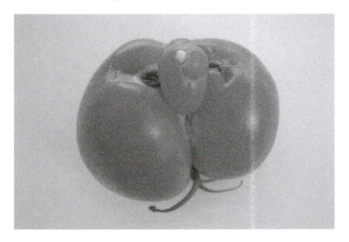

FIGURE 5.15   Catfacing of tomato.

have time to correct the problem before the plant becomes much stressed causing reductions in its yields.

While we all hope that we will not encounter such difficulties, these are natural occurrences that keep us learning about caring for our plants.

6 Water Culture (True Hydroponic Systems)

HYDROPONICS VS. SOILLESS CULTURE

By definition *hydroponics* means "hydro" (water) and "ponos" (working), literally "water working." In a broad sense hydroponics means growing plants without the use of soil, but with or without some inert substrate. Hence, it can also be termed "soilless culture." True hydroponic culture then would be a water culture system without the use of a medium. Today the most common methods of true hydroponic culture are nutrient film technique (NFT) and water culture such as the raft culture system. When a nutrient solution is applied to plant roots from below as a mist this culture is termed "aeroponics." It is also a true hydroponic system.

The most common form of hydroponic culture today is really soilless culture. However, since the same principles of nutrition apply as with true hydroponics most people never differentiate between soilless culture and hydroponics. Many plants, especially longer cropping ones such as tomatoes, eggplants, peppers, and cucumbers, prefer some form of soilless culture over water culture. There are a number of important factors in choosing the best medium for specific plants. The medium must be sterile, that is, free of pests and diseases, have good water retention but not excessive, good oxygenation for plant roots, and retain its structural integrity so as not to compact resulting in insufficient oxygen for the roots of plants. Rockwool and coco coir are the two most popular substrates today. However, depending upon what is available in your area other suitable substrates include peat, rice hulls, perlite, sawdust, bark chips, vermiculite, sand, gravel, haydite (porous shale), leca (light expanded clay aggregate) (clay pellets), and volcanic rock (pumice). Often mixtures of peat, peatlite, sand, coco coir, and rice hulls make a good medium.

WATER CULTURE SYSTEMS

At present NFT is the most popular water culture system. However, large commercial operations growing lettuce are using the raft culture system. The principle of NFT is to have a shallow flow of water under the roots. The flow must be constant and thin to oxygenate the roots of the plants.

If the flow rate is too slow or the film very deep (no longer a film), oxygen demand of the plant roots will not be met and the roots will start to die. Ideally, the upper portion of the root mat that forms in an NFT channel should be above the water level and exposed to air at 100% relative humidity.

Both the slope and flow rate of the solution affect the dissolved oxygen content of the nutrient solution. As a root mat forms in the NFT channel a "damming" effect of the root mat can lead to stagnation of the nutrient solution with regions of low dissolved oxygen. To keep the flow of solution from getting stagnant use a slope of at least 1 in 50 (2%) or up to 3% to 4% with the NFT channels. It has also been found that flow rates of 2 to 3 liters per minute or slightly greater increase oxygenation. The height of free fall of the solution back to the nutrient reservoir is important in aeration of the solution. More height causes a finer break up of the solution and introduces air bubbles into the tank.

You can also use an air pump with air stones in your nutrient tank to enhance the dissolved oxygen level of the solution. These factors are important in the design or choice of an NFT system whether you build it yourself or purchase a complete ready-to-go system.

CHOOSING THE SYSTEM

The correct choice of system will improve your success in growing. Ask yourself some of these questions. What crops do I want to grow? How many plants of each crop do I wish to grow? What available floor area do I wish to use? Remember the spacing needs of plants discussed earlier in Chapter 3. You must base the number of plants on the total floor area you have available. Do you want a large system or a series of smaller ones? As pointed out earlier do not grow lettuce together with vine crops, as it will get insufficient light. Keep low profile plants in separate systems from the vine crops. Do you want to utilize your hydroponic units most effectively by starting your seedlings in another system of trays with rockwool cubes and blocks? This gives you a head start on the growing cycle by transplanting older plants.

To help you answer some of these questions consider this. Determine how much salad crops you need for your family. How many heads of lettuce a week, and how many pounds of cucumbers, tomatoes, and peppers do you consume weekly? Once you have these numbers and types of crops you eat weekly, you can calculate the total number needed of each. Remember that you need to sow lettuce at least two to three times a week and the time for them to mature may take up to 40 days in the hydroponic system.

Lettuce will grow very well in an NFT system, so that should be an obvious choice for it. However, the vine crops would do better in a soilless system. As a result, your choice of the system for your lettuce will differ from that of the vine crops.

NUTRIENT FILM TECHNIQUE (NFT) SYSTEMS

There are a number of manufacturers of NFT systems. My purpose here is to show you some of the units available and discuss their differences. Although price ranges are given, these will fluctuate with the different manufacturers and models. American Hydroponics manufactures five models of NFT systems. The regular NFT Gully Kits are of dimensions 4 × 4 or 4 × 6 feet, and are 21 inches tall (Figure 6.1). The NFT channels for lettuce are 2 inches high by 4 inches wide with 1-inch holes to fit five oasis or rockwool cubes (Figure 6.2). The bottoms of the channels have a number of ridges to make the flow of nutrient solution spread across the channel. This prevents dry spots where plants may not get any solution. The kit includes 5 gullies, plumbing galvanized steel table frame, submersible pump, and 30-gallon nutrient reservoir with cover for the price of about $750. The 5 gullies have holes for 25 plants of lettuce, basil, or bunches of herbs. This spacing of approximately 7 to 8 inches between plants is optimum for these plants.

FIGURE 6.1 NFT Gully Kit growing different lettuces. (Courtesy of American Hydroponics, Arcata, California.)

FIGURE 6.2   Cross-section of NFT channel. (Courtesy of American Hydroponics, Arcata, California.)

In the past American Hydroponics had a NFT Rockwool Gully Kit that came in the same table dimensions as the aforementioned one. The difference was that the NFT channels were 3 inches high by 6 inches wide and therefore better adapted to growing vine crops like tomatoes, peppers, eggplants, and cucumbers. The gullies had larger 3-inch square holes to hold five 3-inch rockwool blocks (Figure 6.3). This kit came complete as described for the NFT Kit for a comparable price to the regular NFT Gully Kit. Although the 4 × 4 foot kit had 23 holes for plants, we know that this number of vine crops could not be grown so close together, even if they were spacing out the tops in an espalier form. The maximum number of tomatoes, peppers or eggplants that could be grown in this area of the table is four to five plants or about two to three cucumbers. At present this kit is no longer available and American Hydroponics has substituted another line of Ebb & Flood Econo System, one- and two-tray units with 9 and 18 square feet of growing space, respectively. These will be discussed later in the section on ebb and flow.

American Hydroponics also in the past had an NFT Wall Garden, which is like a lean-to arrangement of channels. This is specifically for lettuce and herbs. Such a system could be set up yourself on a fence, wall, garage, or any vertical surface where there is good light, as shown in Figure 6.4 and Figure 6.5. Simply fasten some NFT channels to a wall and slope one

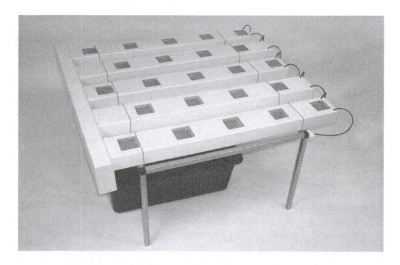

FIGURE 6.3   NFT Rockwool Gully Kit by American Hydroponics. (Courtesy of American Hydroponics, Arcata, California.)

FIGURE 6.4   NFT Wall Garden by American Hydroponics growing various types of lettuce. (Courtesy of American Hydroponics, Arcata, California.)

FIGURE 6.5   NFT Wall Garden growing strawberries. (Courtesy of American Hydroponics, Arcata, California.)

channel to the next with a drainpipe to the next lower channel and finally a return line to the nutrient reservoir. A submersible pump can circulate the solution to the top channel and drain from the bottom one. Three to four tiers of channels would be enough with about 16 to 18 inches between the NFT channels. This would, of course, be for summer use in your backyard. This system would use the standard 2 × 4 inch gullies.

American Hydroponics also has a 612 Herb/Lettuce NFT System that is designed for a small commercial grower or a hobby grower with a large appetite for lettuce and herbs. This unit measures 6 × 12 feet on a bench 34 to 38 inches high (Figure 6.6). It has eight 2 × 4 inch finishing gullies with 18 holes each for a total of 144 plant sites and two 2 × 4 inch nursery gullies with 144 plant sites, which allow you to rotate new seedlings into the system with less space occupied. The finishing channels have 1¾-inch holes at 8-inch centers and the nursery channels have these same diameter holes at 2-inch centers. For a price of approximately $1500 the kit has the 10 channels, the plumbing, galvanized steel table frame, submersible pump, and a 35-gallon reservoir with cover.

Another company, North American Hydroponics, makes small NFT systems from PVC pipes. They come in various sizes from a single pipe system for 6 to 11 plants (Figure 6.7) for about $400 to larger systems consisting of 6 pipes each holding 11 plant sites (total of 66 plants) (Figure 6.8) for about $1500. The modular design of the system allows any size system to be assembled. The reservoir is the PVC pipe base of the growing channels so when expanding the number of channels, the reservoir is also

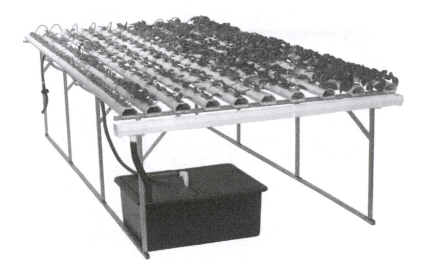

FIGURE 6.6   612 NFT Production Unit measures 6 × 12 feet with nine production channels and one seedling channel. It is ideal for lettuce and herbs. (Courtesy of American Hydroponics, Arcata, California.)

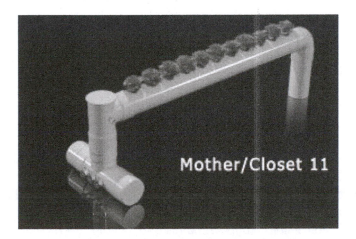

FIGURE 6.7   PVC pipe NFT system by North American Hydroponics. (Courtesy of North American Hydroponics, LLC, Granite Falls, Washington.)

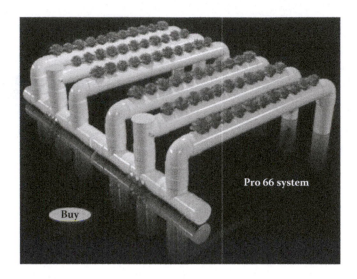

FIGURE 6.8   Pro 66 System of PVC pipe NFT by North American Hydroponics holds 66 plants. (Courtesy of North American Hydroponics, LLC, Granite Falls, Washington.)

increased in size to support the additional plant sites. This system is best for growing lettuce and herbs.

EBB AND FLOW WATER CULTURE SYSTEMS

An ebb and flow system floods the plant roots from below and then drains back to a nutrient reservoir (Figure 6.9). This is not a continuous flow system as is the NFT, but operates on four to eight cycles per day depending upon environmental conditions and plant growth. A submersible pump in a reservoir below periodically floods a bedding tray aerating the plant roots as the solution enters and drains back to the reservoir (Figure 6.10). The plants sit in a tray insert having holes in it to hold either rockwool blocks or growing pots where the plants are seeded or transplanted into a perlite or gravel medium. If rockwool blocks are placed in the plant sites this system is water culture; however, if pots with a substrate are used it is really soilless culture.

American Hydroponics makes several models, the One Tray Econo System and Two Tray Econo System that are priced at $400 to $600. A galvanized steel table frame supports one or two 3 × 3 feet black ABS plastic trays. A 30-gallon reservoir with a cover sits below on a shelf of the frame. A submersible pump in the reservoir pumps up the nutrient solution through a plastic pipe to the flood (bedding) tray. A timer operates the pump. A Gro Well insert accessory is available that is a custom built insert with thirty-six 1½-inch wells designed for use with rockwool cubes.

FIGURE 6.9   The American Hydroponics Vegi-Table is an ebb and flow system that floods the plant roots from below and then drains back to a nutrient reservoir. (Courtesy of American Hydroponics, Arcata, California.)

It nests inside the 3 × 3 foot tray. A drip system can be added to irrigate the pots of substrate or rockwool blocks inserted in the Gro Well option cover. Another insert, a Net Well, can be placed on top of the tray to house twenty-four 5½-inch diameter net pots. This lid also prevents light from entering the plant roots so that algae will not grow in the nutrient solution below. These inserts are adequate for growing lettuce and herbs or seedlings of other crops.

Botanicare, formerly American Agritech, has an ebb and flow unit called Econojet. It consists of a flood tray available in several sizes that is attached to a covered 20-gallon reservoir (Figure 6.11). The trays will hold either rockwool slabs or 6-inch square pots using expanded clay pebbles. Ridges in the flood trays raise the growing pots to keep the growing substrate of the pots out of the nutrient solution providing good aeration to the plant roots. The solution level can be adjusted by a modular fitting as shown in Figure 6.12 for the Micro Garden. Their line of hydrogardens include several different sizes and configurations from a 2 × 2 to a 4 × 8 foot growing space ranging in price from $230 to $1340.

FIGURE 6.10 Healthy roots of plants growing in the ebb and flow tray of the Vegi-Table. Note the fill/drain pipe in the base of the tray on the right. (Courtesy of American Hydroponics, Arcata, California.)

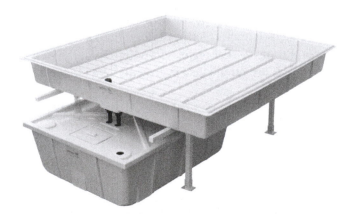

FIGURE 6.11 The Botanicare Econojet ebb and flow system basic tray above a reservoir. This is a 4 × 4 foot unit. (Courtesy of Botanicare, Tempe, Arizona.)

FIGURE 6.12   Ebb and flow fill and drain pipes at the base of the growing tray of the MicroGarden. (Courtesy of Botanicare, Tempe, Arizona.)

Botanicare has a series of Microgardens using ebb and flow systems. Basically, a growing tray sits on top of a nutrient reservoir (Figure 6.13). The solution is pumped into the growing tray with pots of expanded clay pebbles. Large square pots or smaller net pots may be used depending upon the plants to be grown (Figure 6.14). Prices range from $250 to $360.

AEROPONIC SYSTEMS

Botanicare makes an Aerojet system, which is a combination of aeroponic and soilless cultures. The grow trays are 8 inches wide by either 42 or 72 inches long. The 42-inch tray will hold either six 4-inch or four 6-inch mesh pots (Figure 6.15 and Figure 6.16). Mist is applied from below in bursts of 1 minute on and 4 minutes off (Figure 6.17). Fill the pots with expanded clay pebbles. A 20-gallon reservoir comes with the smaller models. Prices range from $800 to $1700 for models with two trays to eight trays.

Botanicare also has a series of smaller aeroponic Microgardens and Turbogardens that use an aeroponic growing tray containing the misters (Figures 6.18 to 6.20). The tray sits on top of a nutrient reservoir and has a cover to support the plants. Prices range from $350 to $550.

FIGURE 6.13   Pots are filled with expanded clay pebbles in an ebb and flow system. (Courtesy of Botanicare, Tempe, Arizona.)

FIGURE 6.14   Smaller net pots may be used instead of the larger square pots to hold the clay pebble substrate. (Courtesy of Botanicare, Tempe, Arizona.)

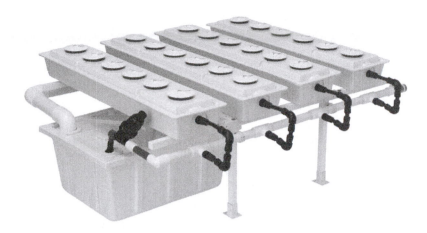

FIGURE 6.15 An Aerojet aeroponic system with four trays of 4-inch pots. (Courtesy of Botanicare, Tempe, Arizona.)

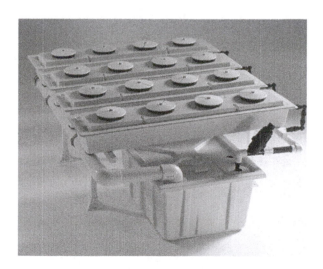

FIGURE 6.16 An Aerojet aeroponic system with four trays of 6-inch pots. (Courtesy of Botanicare, Tempe, Arizona.)

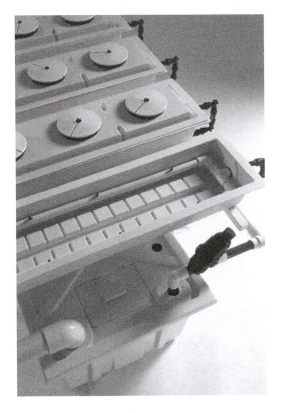

FIGURE 6.17 An Aerojet aeroponic system showing the mist "jets" in the base of the tray of the first tray. Note the nutrient reservoir below and the filter in the inlet line near the first tray. (Courtesy of Botanicare, Tempe, Arizona.)

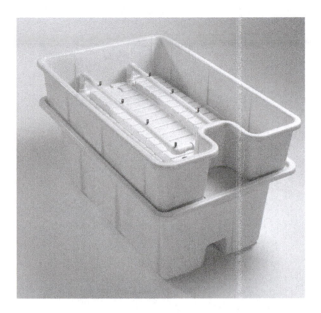

FIGURE 6.18   An aeroponic "MicroGarden" showing the mister jets at the base of the growing tray. (Courtesy of Botanicare, Tempe, Arizona.)

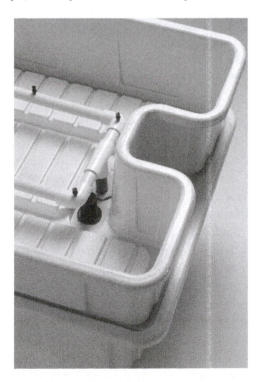

FIGURE 6.19   Close-up view of mist nozzles and plumbing from pump below in the nutrient reservoir of the MicroGarden. (Courtesy of Botanicare, Tempe, Arizona.)

FIGURE 6.20   The aeroponic MicroGarden with the pots and plant support tops in place. (Courtesy of Botanicare, Tempe, Arizona.)

Many of these manufacturers also make propagation units. Most are an aeroponic system. These units are principally for rooting vegetative cuttings. They are particularly suitable to ornamentals. Since we start our vegetable crops from seeds, as was described in Chapter 2, there is no reason to discuss these units. I only want to say that basically most are aeroponic systems such as the RainForest units by General Hydroponics and the Clone Machines of Botanicare. Depending upon their size these units are priced from $240 to $450.

A new product that is an aeroponic system is the Omega Garden. This is a device that supports a slowly rotating growing cylinder that turns as a water pump either mists or submerses the roots with a nutrient solution (Figure 6.21). This rotating garden is called a Volksgarden. The unit measures 48 inches in diameter by 76 inches tall by 30 inches wide, which includes the supporting stand.

Seeds are started in rockwool cubes and transplanted to 3-inch rockwool cubes that are placed into channels of the rotating garden cylinder (Figure 6.22). A motor drives the cylinder that rotates about a central light of 400 to 1000 watts. The cylinder is constructed of ABS food-grade

FIGURE 6.21   Aeroponic rotating system by Omega Garden growing basil. (Courtesy of Omega Garden International, Vancouver, British Columbia, Canada.)

FIGURE 6.22 Rockwool cubes are placed into the channels of the Omega Garden. Volksgarden growing lettuce. (Courtesy of Omega Garden International, Vancouver, British Columbia, Canada.)

plastic. It will accommodate up to 80 plants. The cylinder is chain driven to rotate 24 hours with one rotation per 45 minutes. Very productive crops can be grown with this unit, especially low-profile plants like lettuce, basil, chard, and herbs (Figure 6.23).

A submersible pump is placed in a 50- to 100-gallon plastic reservoir below the rotating garden. The pump is connected to a 24-hour timer that activates one irrigation cycle several times a day (depending upon plant growth) for the full rotation of about 45 minutes per irrigation cycle. When the pump shuts off the water flows back from the reservoir tray directly under the bottom of the rotating cylinder to the tank below through the pump. On an irrigation cycle the pump in the nutrient solution tank fills the upper reservoir sufficiently to make contact with the rooting medium of a minimum of ¼ inch. This level is regulated by an overflow drain that feeds excess solution back to the lower nutrient tank. As the cylinder rotates it

FIGURE 6.23 Omega Garden single carousel growing chard. Note the orientation of the plants toward the central light. (Courtesy of Omega Garden International, Vancouver, British Columbia, Canada.)

passes the growing blocks with the plant roots through this upper reservoir of nutrient solution. The price of this unit is $2595.

Another product is the AeroGarden. Although its name implies an aeroponic system, it really is a drip system and therefore will be discussed in Chapter 7.

COMBINATION WATER CULTURE AND SOILLESS SYSTEMS

If an NFT or ebb and flow system uses large mesh pots with a substrate such as perlite or pebbles, it really is not a true water culture system but a combination of water and soilless cultures. If the mesh pots are only 2 inches in diameter, the system could be termed water culture, as the majority of the plant roots would be in the solution below. But, with larger pots, more roots should remain in the substrate and therefore in my opinion it is a combination of soilless and water cultures. This does not mean that the system is less functional; in fact, it may be more suitable to a wider range of plants, especially for vine crops. Vine crops would do best in the larger pots of 5½ or 8 inches diameter.

Small units of this combination system are available from several companies such as American Hydroponics, American Agritech (Botanicare), and General Hydroponics.

FIGURE 6.24 Baby Bloomer ebb and flow unit growing dwarf patio cherry tomatoes, basil, and lettuce. (Courtesy of American Hydroponics, Arcata, California.)

A small ebb and flow unit is the Baby Bloomer by American Hydroponics with dimensions of 31 × 14 × 12 inches at a price of between $200 and $250. This is a self-contained unit with a reservoir underneath supporting a bedding tray where the solution is pumped periodically to flood the base of the mesh pots containing the plant roots (Figure 6.24). A tray that sits on top of the bedding tray supports 10 mesh pots having perlite or rocks as a medium. Although this unit is fairly small, it would grow a combination of lettuce, herbs and perhaps several small "patio" type cherry tomato plants.

American Hydroponics has several larger units called One and Two Tray Econo Systems, which are the same as was described in the earlier section on ebb and flow. The difference is that these units use 23 mesh grow pots with an expanded clay medium per 3 × 3 foot growing tray instead of the 36 sites for rockwool cubes. At that spacing you could grow lettuce, herbs, and a few vine crops, provided you train them correctly as described earlier. Prices are comparable to those discussed in the ebb and flow section at $400 to $600.

A small ebb and flow system by General Hydroponics has six channels sitting on top of a nutrient reservoir (Figure 6.25). It holds up to 42 plants such as lettuce, arugula, and spinach (Figure 6.26).

General Hydroponics presents a somewhat different design in a combination of NFT and rock culture. It has three models (20, 30, and 60)

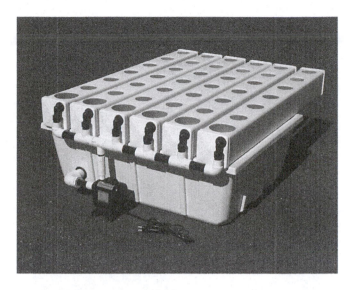

FIGURE 6.25   AeroFlo ebb and flow system by General Hydroponics. (Courtesy of General Hydroponics, Sebastopol, California.)

FIGURE 6.26   This small AeroFlo unit grows 42 low profile plants such as lettuce (lower left), arugula (center), bok choy, and spinach. (Courtesy of General Hydroponics, Sebastopol, California.)

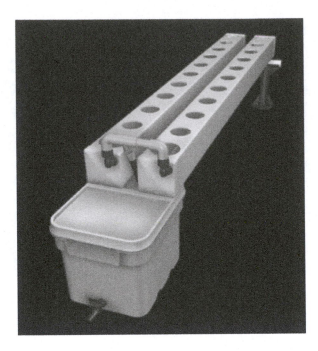

FIGURE 6.27 The AeroFlo^2 20 has two 6-foot growing chambers that hold 20 plants in 3-inch diameter net pots containing clay pellets or coco coir. (Courtesy of General Hydroponics, Sebastopol, California.)

of what it calls AeroFlo^2 20, a 20-site system, as shown in Figure 6.27. It includes an 8-gallon reservoir, two 6-foot grow chambers, a pump, piping, support structure, 3-inch mesh pots, and coco coir or GroRox clay pellets (Figure 6.28). The grow chambers are approximately 6 × 6 inches so they hold up to 4 gallons of nutrient solution each making the total capacity of solution 16 gallons with the drain overflow tubes set at their high position. If the tubes are set all the way down to the base of the grow chamber, then all the solution will be retained by the reservoir only to give a total of 8 gallons. This gives the system the capability to function as deep flow with the drain tubes set high and as an NFT system with the tubes set low to the bottom of the grow chambers. They suggest using the high position when transplanting and lowering the solution level to an NFT system once the plant roots emerge into the grow chamber. The suggested retail price of this unit is $350.

The AeroFlo^2 30 site system differs from the smaller 20 unit in that it has three 6-foot grow chambers with a 20-gallon reservoir and uses thirty 3-inch mesh pots for the plants (Figure 6.29). The total reservoir capacity in the flooded stage (overflow tubes set high) is approximately 34 gallons. The model 60-site system having six grow chambers may be configured

FIGURE 6.28 GroRox clay pellets. (Courtesy of General Hydroponics, Sebastopol, California.)

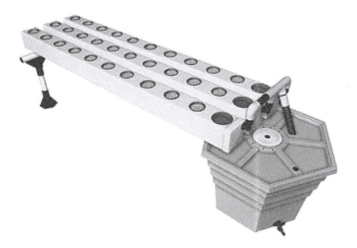

FIGURE 6.29 The AeroFlo^2 30 holds 30 plants in its three 6-foot growing trays. (Courtesy of General Hydroponics, Sebastopol, California.)

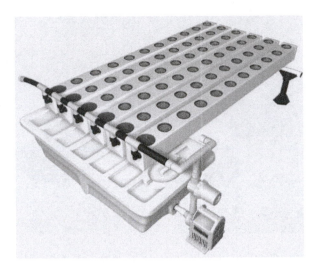

FIGURE 6.30 The AeroFlo^2 60 may be arranged with the six growing trays on one side of the reservoir for smaller plants as herbs and lettuce. (Courtesy of General Hydroponics, Sebastopol, California.)

in two ways; the entire six chambers on one side for small crops such as lettuce and herbs (Figure 6.30 and Figure 6.31) or three chambers on each side of the 24-gallon reservoir for larger crops such as peppers and tomatoes (Figure 6.32). The total system capacity in the flooded stage is 45 gallons. Suggested retail prices for these units are $475 and $915, respectively. Two of the 60-site system units may be connected to form 120 plant sites. This extension module costs about $750.

The Econojet system by Botanicare as described in the ebb and flow section can also be considered a combination system when using twelve 6-inch pots with expanded clay pebbles instead of the rockwool blocks as mentioned earlier.

FIGURE 6.31 A nice crop of lettuce growing in the AeroFlo^2 60 system. (Courtesy of General Hydroponics, Sebastopol, California.)

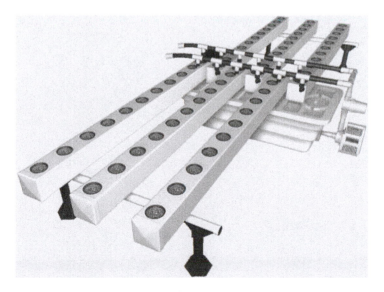

FIGURE 6.32 Alternatively the AeroFlo^2 60 may have three trays on each side of the reservoir for taller plants such as tomatoes. (Courtesy of General Hydroponics, Sebastopol, California.)

7 Soilless Culture

SUBSTRATES (MEDIA)

The most popular substrates for small hydroponic units are expanded clay pebbles, rockwool, and perlite. All substrates must be inert, provide adequate oxygenation, and yet have good water retention as described earlier (Chapter 2). On a small scale we do not sterilize the medium and reuse it, as that is somewhat difficult and the amount of medium required per cropping period is generally fairly small and inexpensive. Many of the small commercially available hydroponic units that use mesh (net) pots are most suited to expanded clay pebbles, other small rocks, or perlite. Rockwool slabs to grow mature plants will have to fit into narrow, long trays as each slab is 6 to 8 inches wide by 3 to 4 inches thick by 36 inches long.

SOILLESS SYSTEMS–EXPANDED CLAY

Most combination water and soilless culture systems use expanded clay pebbles as has already been described. Various systems of soilless cultures have been designed by a number of manufacturers. Starting with the smaller one-pot systems of General Hydroponics, it has a WaterFarm (Figure 7.1 and Figure 7.2), a PowerGrower (Figure 7.3), and an EcoGrower (Figure 7.4). They differ mainly in size of growing pots and reservoirs. The WaterFarm uses a 2-gallon grow pot and a 4-gallon reservoir, whereas the PowerGrower has a 3-gallon grow pot and 5.7-gallon nutrient tank. The EcoGrower has a 17-gallon reservoir and with a drip system to each of six 6-inch mesh pots for growing larger plants. The growing pot(s) filled with expanded clay pebbles (Grorox) sit on top of the reservoir container. The nutrient solution from the reservoir moves up by an air pump positioned outside and attached to a tube that enters a small piece of PVC pipe and is connected to a circular perforated drip ring or spider drip assembly in the case of the EcoGrower. This system is best suited to herbs or a single plant like a dwarf patio tomato plant. They are priced from about $50 to almost $200 for the EcoGrower.

The PowerGrower may be connected in a series of pots (Figure 7.5). This 8-Pak Kit is a top feed, drip system with seven 3-gallon growing chambers. The nutrient reservoirs are termed a "Standard Controller," which is a 13-gallon master reservoir positioned on top of a float-activated

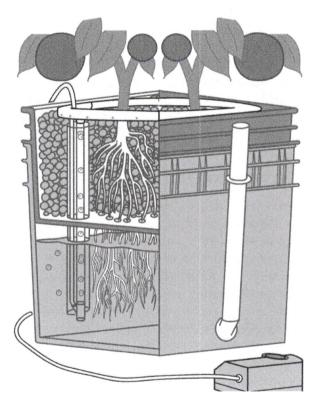

FIGURE 7.1   Diagram of WaterFarm one-pot system of General Hydroponics. (Courtesy of General Hydroponics, Sebastopol, California.)

FIGURE 7.2   WaterFarm components include pot, reservoir, pump, tubing and clay pellet substrate. (Courtesy of General Hydroponics, Sebastopol, California.)

FIGURE 7.3   The PowerGrower is a stand-alone top feed/drip system. It is best suited to a larger plant such as a tomato. (Courtesy of General Hydroponics, Sebastopol, California.)

FIGURE 7.4   The EcoGrower has an air-driven drip system. It can accommodate six 6-inch net pots. (Courtesy of General Hydroponics, Sebastopol, California.)

FIGURE 7.5   The PowerGrower may be connected in a series of pots such as this 8-Pak Kit. (Courtesy of General Hydroponics, Sebastopol, California.)

8-gallon reservoir. As solution is used up by the plants from the lower reservoir, an automatic float valve will draw solution from the upper master reservoir. This system is fine for growing vine crops such as tomatoes and peppers. The nutrient controller system costs about $70 and a complete 8-Pak system costs $525.

General Hydroponics has a somewhat larger unit called EuroGrower. This system consists of eight plastic Dutch Bato Bucket pots (Figure 7.6) that sit on top of a 40-gallon reservoir. The dimensions are 4 feet 7 inches long, 2 feet wide, and 2 feet tall. Each Bato bucket has a siphon drain fitting that maintains a water level of about 1 inch at the bottom of the pot. It drains the pots to the underlying reservoir. A submersible pump in the reservoir feeds the plants with a drip irrigation line to each pot. The pots are filled with a mixture of Grorox and coco coir.

Vegetables and flowers grow well in this system. It is priced at $525. There are also larger systems for growing vine crops such as tomatoes (Figure 7.7).

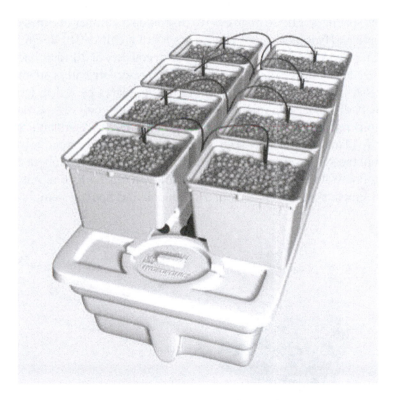

FIGURE 7.6 The "EuroGrower" consists of 8 Bato buckets of 8 gallons each sitting on top of a 40-gallon reservoir. (Courtesy of General Hydroponics, Sebastopol, California.)

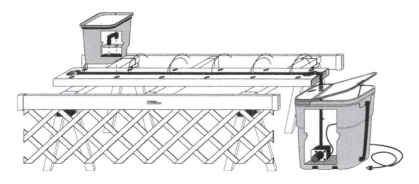

FIGURE 7.7 A recirculating vine crop system for 12 Bato buckets. (Courtesy of General Hydroponics, Sebastopol, California.)

I have good success with growing tomatoes, peppers, eggplants, and cucumbers commercially in Bato buckets with a perlite substrate. They all grow very well with perlite, so I think that it would be better to use perlite instead of the expanded clay pebbles mixed with coco coir. Just be careful with the spacing for these plants as we discussed a number of times earlier.

American Hydroponics has a similar system it calls a 210 Recirculating Vine Crop System (Figure 7.8). This system consists of 10 Bato buckets, a galvanized steel table frame, and a 35-gallon reservoir with a submersible pump that supplies the nutrient solution to the plants by a drip irrigation system. Perlite alone or a mixture of perlite and coco coir (75% perlite, 25% coco coir) makes a suitable substrate. Twenty tomatoes, eggplants, or peppers (two plants per pot) or 10 European cucumbers (1 plant per pot) can be grown in the system. This is a recirculating system and costs about $1495.

AutoPot Watering Systems, based in the United Kingdom, has developed a simple system that is similar to an ebb and flood system; however,

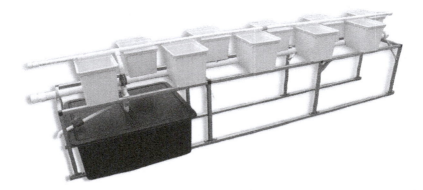

FIGURE 7.8 The 210 Recirculating Vine Crop System by American Hydroponics. (Courtesy of American Hydroponics, Arcata, California.)

FIGURE 7.9 A 1Pot System. (Courtesy of AutoPot Global, Ltd., Paddington, Oxfordshire, United Kingdom.)

the unique difference of the AutoPot systems is that it is the only system that allows the plants to be in control of their own requirements. Their AQUAvalve is the key to the AutoPot Watering System. It controls the flow of water or nutrient solution to the plants by gravity pressure from a tank/reservoir without the use of pumps, timers, and power. Each 4-gallon growing pot sits in a tray beside the AQUAvalve that regulates the water level to 2 cm (¾ inch) (Figure 7.9). Once the water level of 2 cm is reached, the AQUAvalve will shut off the incoming water resulting in no more water being allowed to enter the tray via the AQUAvalve until the plant has consumed all the water that was previously delivered. The water level completely drains before any further water is allowed to enter.

The growing pots are connected to a tank/reservoir by a network of poly tube irrigation lines. There are numerous substrates that can be used; one that is recommended is a mixture of 1:1 coir to perlite.

The system can be expanded to up to 48 pots for small growers (Figure 7.10). AutoPot Watering Systems has two types of pot growing systems: the 1Pot module shown in the photos and the easy2grow system. Both designs can be used to create large multipot systems. In the hobby market, both designs are used generally up to 100 pots. For the commercial growers, using thousands of pots, the system of choice is the easy2grow extension kit. For commercial growers at present, AutoPot has installed systems of over 2000 pots. AutoPot claims that there is no limit to the size of the growing system; it is simply a matter of adding more pots, trays, and larger tanks.

Prices for the 1Pot System start at $55, $80 for the 2Pot System, $160 for the 4Pot System, $325 for the 8Pot System, $470 for the 12Pot System, $900 for the 24Pot System, and $1415 for the 48Pot System. The prices for

FIGURE 7.10   24Pot System. (Courtesy of AutoPot Global, Ltd., Paddington, Oxfordshire, United Kingdom.)

the easy2grow Systems for the same sizes are $50, $75, $100, $275, $520, and $900.

AEROGARDEN

Although the name "AeroGarden" suggests an aeroponic system, it is actually a drip irrigation system using small foam growing cubes that are termed "Growing Sponges." The AeroGarden, produced by AeroGrow in Boulder, Colorado, was first introduced to the market in 2005. The AeroGarden is a plug-and-play type of unit. The unit has its own lights and the kit comes with seeds, nutrients, and an instruction manual (Figure 7.11 and Figure 7.12). The philosophy of the AeroGarden is to make hydroponic growing available to anyone and to be done on your kitchen counter. The growing of the plants is automated with a light timer and irrigation cycle controller all built into the unit. Over 40 types of seed kits are available including combinations of flowers, patio cherry tomatoes, dwarf peppers, lettuces, mesclun mixes, basils, and herbs.

There are now a number of models of the AeroGarden. The initial Classic 7 has seven growing sites and the AeroGarden 3 has three growing sites. There is also an AeroGarden Extra Tall for larger tomato and pepper plants. The AeroGarden 3 is also available in a number of colors they term "Expressions," in addition to the standard black, white, and silver finishes

FIGURE 7.11 AeroGarden nutrients and seed kits. (Courtesy of AeroGrow, Boulder, Colorado.)

FIGURE 7.12 The AeroGarden Classic 7 in black with seven growing sites. (Courtesy of AeroGrow, Boulder, Colorado.)

FIGURE 7.13 The AeroGarden 3-Ladybug has three plant growing sites. (Courtesy of AeroGrow, Boulder, Colorado.)

of all of the units (Figure 7.13 and Figure 7.14). With their unique elegant modern look it makes them very attractive as a display in your kitchen. The prices vary from $90, $150, to $200 for the smaller to the larger models. Seed kits cost from $12.95 to $17.95.

ROCKWOOL AND COCO COIR CULTURES

With rockwool or coco coir culture start the seedlings in rockwool cubes and transplant to rockwool blocks before transplanting a second time to the rockwool or coco coir slabs. Setting up your own rockwool/coco coir system is fairly simple. You need a 20- to 30-gallon solution tank, support structure, pump and drip irrigation lines, PVC piping, and a series of trays in which to place the rockwool or coco coir slabs. The trays should be at least 8 inches wide by a multiple of 3 feet as the slabs are approximately 3 feet long.

Botanicare sells what they term "plumbed grow trays" for rockwool or ebb and flow systems. They have 44 × 6 × 4 inch and 42 × 8 × 4 inch trays complete with plumbing fittings for drainage and drip irrigation. These hold one slab or six 6-inch square pots. Botanicare also has 44 × 6 × 4 inch and 42 × 8 × 4 inch ebb and flow trays. Purchase these trays for your system from $20 to $30 each and build the rest. You can buy a small storage

FIGURE 7.14 The AeroGarden Extra Tall for taller plants such as tomatoes. (Courtesy of AeroGrow, Boulder, Colorado.)

box (available in Kmart, Target, or Walmart) to serve as the nutrient reservoir. Pumps and irrigation supplies are available in many garden centers, hydroponic shops, and irrigation stores.

Alternatively, you may purchase complete systems such as those available from Botanicare. Botanicare has a number of models under its Jetstream category. Two-tray, four-tray, six-tray and eight-tray units are available in 6, 8, or 9 inches wide by 42 to 44 inches long (Figure 7.15 and Figure 7.16). Each tray contains one rockwool or coco coir slab. A submersible pump feeds the individual plants in their rockwool blocks with a drip irrigation line via a PVC header. A 2-inch diameter PVC catchment pipe returns the solution to the 20- or 30-gallon reservoir. Four plants are set onto each slab in each grow tray. A four-tray system has 16 plant sites and a six-tray one has 24 sites, but always remember the spacing rule. This rockwool or coco coir system is ideal for most vine crops, just be careful with the area per plant needed. Prices range from $600 for the two-tray unit to over $1000 for an eight-tray one. These units are easily converted to mesh pots with expanded clay pebbles or perlite using the same system of drip irrigation (Figure 7.15).

Botanicare also sells a one-tray and two-tray system it calls a Jetflo that is an ebb and flow system. The one-tray unit consists of one tray 7 inches

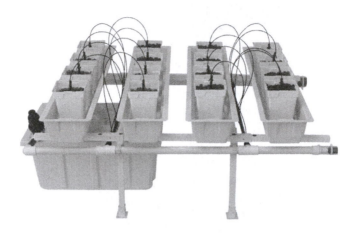

FIGURE 7.15   The four-tray Jetstream with 6-inch pots, clay pebbles, and drip irrigation system. Lower photo shows details of the drainage ridges to provide good drainage to the pots or slabs. (Courtesy of Botanicare, Tempe, Arizona.)

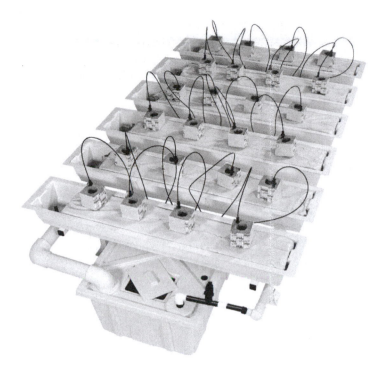

FIGURE 7.16   The six-tray Jetstream with rockwool blocks and slabs and drip irrigation system. (Courtesy of Botanicare, Tempe, Arizona.)

deep by 2 feet by 4 feet that sits on a bench delivery and drainage module. The system includes a pump, reservoir, and bench. The two-tray unit is very similar but has two trays instead of one. The prices for these two units are from $400 to $500.

PERLITE CULTURE–VERTICAL PLANT TOWERS

Plant towers are a great way to increase your production in a limited area of space. In my experience lettuce, bok choy, herbs, and strawberries do well in plant towers. The substrate of choice is perlite. You may wish to mix 15% to 25% coco coir with the perlite to increase its water retention. However, the more coco coir in the mix, the less frequent and duration of irrigation cycles will be necessary as the mixture retains more moisture than perlite itself.

Verti-Gro in Florida developed the unique system of vertical culture using Styrofoam pots stacked one on top of the other up to 8 to 10 pots high. The culture was initially developed for the growing of strawberries in greenhouses. With these towers the plant density can be increased up to 8

FIGURE 7.17   Plant tower formed by stacking Styrofoam pots one on top of the other by turning them 45 degrees to each other. Notches fit each pot in its exact position. (Courtesy of CuisinArt Golf Resort & Spa, Anguilla.)

times of what would be possible when growing them in normal horizontal beds. Verti-Gro soon became aware that low-profile crops such as spinach, herbs, and lettuce adapted readily to this system of growing.

The Styrofoam pots are stacked one above the other by rotating them by 45 degrees so that the four corners of each pot are exposed, as shown in Figure 7.17. The Styrofoam pots measure 9 × 9 × 8 inches with slightly tapered sides. Their top lip is notched on each side to fit the bottom of the adjoining pot (Figure 7.17). Here is how they are set up. A collection pan sits on a drainpipe or supported at that level to collect the drain water and return it to a solution reservoir. An indoor unit for your house would need to be raised on a support frame to keep the drainpipe at an elevation above the level of the nutrient reservoir. The drain pot or pan should connect to the drainage return pipe by a flexible tube or hose line. A ¾-inch diameter PVC tee can be placed at the bottom of the drain pot. A piece of galvanized steel electrical conduit pipe is placed into the upper end of the tee. A ¾-inch diameter PVC pipe slides over the conduit to act as a sleeve over which the pots are threaded through a central hole in each pot (Figure 7.18). A swivel

FIGURE 7.18 Electrical conduit with plastic pipe sleeve supports pots vertically from an overhead wire. Various herbs growing in plant towers of 10 pots. (Courtesy of CuisinArt Golf Resort & Spa, Anguilla.)

plate between the top of the tee extension and the first pot allows the entire tower to be rotated every day to better distribute the light entering the crop. The swivel plate is constructed from ½ inch thick plastic and has a hole large enough to permit the conduit to go through it into the ¾-inch tee below sitting in the drainage pot.

In your house you will probably only be able to stack eight pots high, as space is required for the lights above. Slide all the pots onto the vertical pipe support before filling them with the perlite or other substrate. Then, level the support pipe in all directions before attaching the top of it to a support wire strung across the room. Once the pots are supported on the central pipe, fill each pot starting from the bottom one with perlite up to within 1½ inches of the rim. Allowing a bit of air space between each pot will help to prevent roots growing from one pot down into the next through the perforated bottoms of the pots. The next step is to install a drip irrigation system.

Using a submersible pump in a nutrient reservoir, attach a PVC pipe vertically to the upper height of the tower and then convert it to ½-inch black

FIGURE 7.19   Two drip lines enter the top pot and a third one is located in the middle of the plant towers with thyme. (Courtesy of CuisinArt Golf Resort & Spa, Anguilla.)

polyethylene hose running along the support wire to all towers. If you put a ¾-inch tee on top of the ¾-inch PVC sleeve of each tower, you can run the black poly line through these tees to support the poly line. Insert two drip lines to the top pot and one halfway down the tower. This will assist in distributing the solution evenly down the plant towers (Figure 7.19). Each plant tower requires 10 square feet of floor area to permit adequate light penetration to the crop.

These towers are very productive (Figure 7.20). One tower of eight pots will produce 32 heads of lettuce or bok choy monthly if you start the seedlings in rockwool or Oasis cubes, then transplant them into the plant tower at 26 to 28 days as they take about 60 days from seed to maturity (Figure 7.21). Sequence the planting so that you may harvest every day or so depending upon your personal needs. Herbs will take several months to become well established (Figure 7.20). After that you may cut them daily, as a little trimming each day is better for their continued production than

FIGURE 7.20 Left to right: mint, sage, sweet marjoram, thyme, and chives in plant towers. (Courtesy of CuisinArt Golf Resort & Spa, Anguilla.)

to cut them back severely once a week or so. To grow strawberries you need to purchase live plants as was described in Chapter 2. They will take several months to start producing. Growing vine crops in plant towers is not very convenient as only one plant can be grown in a single pot near the bottom of the tower. If you put many plants in the plant tower, even small bush varieties the top plants will bear well, but the ones in the middle and lower of the tower will not get sufficient light to be productive. We have already talked about many more suitable systems for growing your tomatoes, peppers, eggplants, and cucumbers.

POPULAR HYDROPONICS

"Popular hydroponics" is a term coined by a group of nonprofit organizations from the universities of Colombia and Peru to get people involved in producing their own vegetables in poor communities. The universities together with community members developed very inexpensive, simple, nonmechanical methods of growing hydroponically on a small scale. The principle is to use inexpensive local materials or waste such as old tires or plastic containers. Drainage holes are drilled in the containers prior to filling them with media such as peat, rocks, rice hulls, coco coir, or mixtures of these substrates. Seeds are sown in these gardens and the plants fed with a hydroponic nutrient solution developed by the universities. The growers are given seeds, nutrients, and other supplies at a very low cost. Lettuce,

FIGURE 7.21   Bok choy 50 days from sowing, almost ready to harvest. (Courtesy of CuisinArt Golf Resort & Spa, Anguilla.)

beets, chard, strawberries, herbs, basil, peppers, tomatoes, and local crops of their normal diet are grown (Figure 7.22). In addition, they grow grasses as fodder for their animals (Figure 7.23).

They also use simple systems of column culture with plastic pipes or polyethylene sacks containing a mixture of rice hulls and coco coir. A container with the nutrient solution is attached to the top of the sack or column where it percolates slowly into it. The drainage runs into a collection pan at the bottom so that the solution may be poured back into the container at the top for use again. Strawberries are popular in this system. You can construct a simple raft culture system from lumber, polyethylene, and Styrofoam. Many poor communities in Latin America construct such units to grow lettuce, basil, herbs, chard, spinach, and other leafy crops (Figure 7.24). Cut a 4 × 8 foot Styrofoam sheet in half to get exactly 4 × 4 feet. Cut sixty-four ¾-inch diameter holes at 6 × 6 inch centers with a hole saw for the plant sites. The Styrofoam panel supports the plants and insulates the solution below in the reservoir from light and heat. Make a wooden or brick frame at least 1-inch wider in the inside dimensions (49

FIGURE 7.22 Inexpensive sack culture of growing lettuce, herbs, and strawberries in poor communities of Peru.

FIGURE 7.23 Grasses growing for cattle in inexpensive containerized hydroponics containing mixtures of rock, sand, peat, rice hulls, and coco coir. This is "popular hydroponics" in Peru.

× 49 inches). This gives you room for the Styrofoam to be easily taken in and out of the reservoir. The height should be about 7 to 8 inches. Line the reservoir with double 10 mil polyethylene or a 20 mil swimming pool vinyl. If you have time to agitate the nutrient solution with a whisk three to four times a day, you can get by without an aeration pump. The easiest way to aerate the solution is to use an aquarium pump attached to air stones in the reservoir. This do-it-yourself project should cost under $50.

Although I present these systems here as all manually operated, you can use the same principles to construct such systems and automate them with a tank, submersible pump, and irrigation components. The other point I wish to make here is that hydroponics can be practical for many different system designs from sophisticated, automated ones to simple, manual ones. So it is applicable to all income levels and backgrounds. You do not need to be a rocket scientist to successfully grow hydroponically; just be able to follow instructions to best achieve the results you want for your particular level of growing. These simple systems used for poor communities may be the answer for many of these people currently under malnutrition to better provide basic nutrition for their families. They often live in arid regions where the soils are very poor in nutrition and structure compounded with scarcity of water. In this situation hydroponics is the answer to growing fresh vegetables.

FIGURE 7.24 Inexpensive raft culture system for growing lettuce and herbs in poor communities of Peru.

8 Hobby Hydroponic Supplies and Information

HYDROPONIC SUPPLIES

There are now thousands of hydroponic stores worldwide. Many are listed on the Internet with their Web sites. Also, related industries such as hardware and builders' supplies, irrigation suppliers, garden centers, fertilizer companies, and discount stores sell components needed, especially for the do-it-yourself hobbyist.

Most of these stores have their own Web sites that you can browse to explore what products they offer. In addition, there are the Web sites of some of the manufacturers and distributors of hydroponic equipment. The following Web sites in Table 8.1 include the manufacturers and some distributors of units described earlier in this book. The table lists but a few of the many hydroponic manufacturers and distributors in the United States and Canada. For further companies please refer to the Internet search engines or hydroponic magazines such as "Maximum Yield Gardening" that lists many distributors at the back of their monthly magazine.

HYDROPONIC SERVICES

A lot of information is available through private and governmental Web sites. Such Web sites offer information on pest and disease identification and control, cultural techniques for crops, nutrition, nutritional disorders and remedies, hydroponic systems, seeds, and new cultivars most suitable to hydroponic growing. Some of these sites are presented in Table 8.2.

These are a few sites offering technical information on horticulture and hydroponics. Most have links to other sites on specific topics. Some are very extensive and give references to their publications, and recommend other journals and articles on related topics.

There are many Web sites on integrated pest management (IPM) that have color photos of pests, diseases, and beneficial insects and microorganisms that control them. Table 8.3 summarizes some of these sites. The sites are useful in identification of any pest problem you may encounter.

TABLE 8.1

Web Sites of Manufacturers and Distributors of Hydroponic Units and Accessories

Company	Web Site
Advanced Nutrients	www.advancednutrients.com
AeroGarden	www.aerogarden.com
American Hydroponics	www.amhydro.com
Apache Tech Inc.	www.apachetechinc.com
Autogrow Systems Ltd.	www.autogrow.com
AutoPot	www.autopot.co.uk
Better Grow Hydroponics	www.bghydro.com
Bluelab Corporation Limited	www.getbluelab.com
Botanicare (American Agritech)	www.botanicare.com
Co2Boost	www.co2boost.com
CropKing, Inc.	www.cropking.com
Current Culture H2O	www.cch2o.com
General Hydroponics	www.generalhydroponics.com
Greentrees Hydroponics	www.hydroponics.net
Grodan	www.grodan101.com
Growco Indoor Garden Supply	www.4hydroponics.com
Hanna Instruments	www.hannainst.com
Homegrown Hydroponics	homegrownhydro.com
Horti-Control	www.horticontrol.com
Hydrodynamics International, Inc.	www.hydrodynamicsintl.com
Hydrofarm Horticultural Products	www.hydrofarm.com
Hydrologic Purifications Systems	www.hydrologicsystems.com
Hydrotek	www.hydrotek.ca
Milwaukee Instruments	www.milwaukeeinstruments.com
Myron L Company	www.myronl.com
Nickel City Wholesale Garden Supply	www.ncwgs.com
North American Hydroponics	www.wearehydro.com
P.L. Light Systems	www.pllight.com
Pulse Instruments	www.pulseinstrument.com
Simply Hydroponics and Organics	www.simplyhydro.com
Solis Tek Digital Ballasts	www.solis-tek.com
Sunburst Hydroponics	www.4hydro.com
Sunleaves Garden Products	www.sunleaves.com
Sunlight Supply, Inc.	www.sunlightsupply.com
Verti-Gro, Inc.	www.vertigro.com

TABLE 8.2
Web Sites of Government Agencies and Universities Offering Information on Hydroponics

Agency/Organization	Web Site
University of Arizona, College of Agriculture	www.ag.arizona.edu/hydroponictomatoes
Cornell University, College of Agriculture and Life Sciences/ Greenhouse Horticulture	www.greenhouse.cornell.edu
International Technical Services	www.greenhouseinfo.com
U.S. Department of Agriculture	www.usda.gov
Ontario Greenhouse Vegetable Growers	www.ontariogreenhouse.com
British Columbia, Ministry of Agriculture	www.agf.gov.bc.ca/cropprot/productguide.htm
BC Greenhouse Growers' Association	www.bcgreenhouse.ca/publications.htm
North Carolina Cooperative Extension	www.ces.ncsu.edu
University of California Statewide Integrated Pest Management (IPM) Program	www.ipm.ucdavis.edu
Mississippi State University Extension Service	www.msucares.com/pubs/

Full descriptions of both the pests and beneficials are presented that can assist you in determining what control measures are available.

Seed catalogs are very informative. They indicate the kind of conditions plants like for their growth. They describe the characteristics of the vegetables, cropping periods, number of seeds per unit weight, and other interesting aspects of all varieties. In addition, it is nice to see what your ideal fruits and vegetables should look like. Seed catalogs also introduce new varieties every year with extensive descriptions of their characteristics. Although these catalogs are available online, it is entertaining to spend some leisure time studying the catalogs in printed form. I have listed some Web sites in Table 8.4 for a few seed houses.

HYDROPONIC ORGANIZATIONS AND INTERNET CHAT CLUBS

There are a number of hydroponic societies, some of which are summarized in Table 8.5, that promote new technology and products. They generally have annual meetings or conferences at which they obtain experts within the field of hydroponics to give presentations. I have found these

TABLE 8.3

Web Sites on Identification and Control of Pests and Diseases Using Integrated Pest Management (IPM)

Organization	Web Site
Koppert Biological Systems	www.koppert.nl
Association of Natural Biocontrol Producers	www.anbp.org
International Technology Services	www.intertechserv.com
Biobest	www.biobest.be
Mycotech Corporation	www.mycotech.com
BioWorks	www.bioworksbiocontrol.com
Cornell University, College of Agriculture and Life Sciences, Department of Entomology	www.biocontrol.entomology.cornell.edu/
University of California Statewide Integrated Pest Management (IPM) Program	www.ipm.ucdavis.edu/
New York State Integrated Pest Management (IPM) Program	www.nysipm.cornell.edu/elements/ghouse. asp
Nature's Control	www.naturescontrol.com/controls.html

TABLE 8.4

Web Sites of Seed Houses

Seed House	Web Site
De Ruiter Seeds Inc.	www.deruiterusa.com
Monsanto Vegetable Seeds	www.monsanto.com
Johnny's Selected Seeds	www.johnnyseeds.com
Ornamental Edibles	www.ornamentaledibles.com
Paramount Seeds Inc.	www.paramountseeds.com
Richters Herbs	www.Richters.com
Stokes Seeds Ltd.	www.stokeseeds.com
Rijk Zwaan USA	www.rijkzwaan.nl

TABLE 8.5

Contact Information for Hydroponic Societies

Organization	Mailing Address	Web Site
Hydroponic Society of America (HSA)	P.O. Box 1183, El Cerrito, CA 94530	www.lisarein.com/ hydroponics
Asociacion Hidroponica Mexicana A.C.		www.hidroponia.org.mx
Australian Hydroponic and Greenhouse Association (AHGA) (renamed Protected Cropping Australia [PCA])	Narrabeen, NSW, 2101, Australia	www.protectedcropping australia.com
Centro de Investigacion de Hidroponia y	Nutricion Mineral, Univ. Nacional Agraria La Molina, Av. La Universidad s/n La Molina, Lima 12, Peru	www.lamolina.edu.pe/ hidroponia
Centro Nacional de Jardineria Corazon Verde in Costa Rica		www.corazonverdecr.com
Encontro Brasileiro de Hidroponia		www.encontrohidroponia .com.br
Singapore Society for Soilless Culture (SSSC)	#13-75, 461 Crawford Lane, Singapore 190461	

very informative and a pleasure to meet people who have been inspired by hydroponics. It is always a learning experience to see new research and products. Most conferences have a hydroponic suppliers' trade show displaying products offered by companies.

There are several hydroponic forums where you may sign up to be part of discussion groups online. You may submit questions for advice from other growers and hydroponic experts. This is also a good method of keeping informed of new products. Send an e-mail to the site to sign up as a member. Some are listed in Table 8.6.

Of course, you can also access a lot of information from Web sites like YouTube, Facebook, and Twitter.

HYDROPONIC MAGAZINES

The popular magazines listed in Table 8.7 are a must for any hydroponic grower. They have both poplar and technical articles. These magazines

TABLE 8.6
Web Sites or E-Mails of Hydroponic Forums

Forum	Web Site or E-Mail
bghydro forums	http://forums.bghydro.com/activity.php
Home Hydro Systems	www.homehydrosystems.com/forum/htm
iVillage GardenWeb	http://forums.gardenweb.com/forums/hydro/
Helpful Gardener Gardening Forum	www.helpfulgardener.com/phpBB2/viewforum.php?f=42

TABLE 8.7
Contact Information of Hydroponic Magazines

Magazine	Mailing Address	Web Site
Practical Hydroponics & Greenhouses	Casper Publications Pty Ltd., P.O. Box 225, Narrabeen, 2101, Australia	www.hydroponics.com.au
The Growing Edge Magazine	P.O. Box 1027, Corvallis, OR 97339	www.growingedge.com
Maximum Yield Gardening	2339 Delinea Place, Nanaimo, BC, Canada V9T 5L9	www.maximumyield.com
The Indoor Gardener Magazine	Green Publications, P.O. Box 52046, Laval, Quebec, Canada H7P 5S1	www.theindoorgardener.ca
deRiego	Revista deRiego, Apdo. Postal 86-200, Mexico, D.F. C.P. 14391	www.revistaderiego.com.mx
Urban Garden Magazine		http://urbangardenmagazine.com

also have extensive advertising by manufacturers and suppliers of hydroponic products to keep you informed of new developments.

REFERENCES

Many books are available on hydroponics. Books are sold by hydroponic stores and garden centers, and on the Internet at online book retailers such as www.amazon.com and www.barnesandnoble.com. The Hydroponic Society of America, The Growing Edge, and Practical Hydroponics & Greenhouses as well as the numerous Web sites of hydroponic suppliers all sell books.

The following are helpful books about hydroponics.

Bridwell, R. 1990. *Hydroponic Gardening*, rev. ed., Woodbridge Press, Santa Barbara, CA.

Cooper, A. 1979. *The ABC of NFT*, Grower Books, London.

Dalton, L. and R. Smith. 1984. *Hydroponic Gardening*, Cobb Horwood Publications, Auckland.

Douglas, J.S. 1984. *Beginner's Guide to Hydroponics*, new ed., Pelham Books, London.

Douglas, J.S. 1985. *Advanced Guide to Hydroponics*, new ed., Pelham Books, London.

Harris, D. 1986. *Hydroponics: The Complete Guide to Gardening Without Soil: A Practical Handbook for Beginners, Hobbyists and Commercial Growers*, New Holland Publishers, London.

Jones, J.B. Jr. 1997. *Plant Nutrition Manual*, CRC Press, Boca Raton, FL.

Jones, J.B. Jr. 1999. *Soil and Plant Analysis*, CRC Press, Boca Raton, FL.

Jones, J.B. Jr. 1999. *Soil Analysis Handbook of Reference Methods*, CRC Press, Boca Raton, FL.

Jones, J.B. Jr. 2001. *Laboratory Guide for Conducting Soil Tests and Plant Analysis*, CRC Press, Boca Raton, FL.

Jones, J.B. Jr. 2004. *Hydroponics: A Practical Guide for the Soilless Grower*, 2nd ed., CRC Press, Boca Raton, FL.

Jones, J.B. Jr. 2007. *Tomato Plant Culture: In the Field, Greenhouse and Home Garden*, 2nd ed., CRC Press, Boca Raton, FL.

Jones, L., P. Beardsley, and C. Beardsley. 1990. *Home Hydroponics ... and How to Do It!*, rev. ed., Crown Publishers, New York, NY.

Kenyon, S. 1992. *Hydroponics for the Home Gardener*, rev. ed., Key Porter Books Ltd. Toronto, Canada.

Malais, M. and W.J. Ravensberg. 1992. *Knowing and Recognizing: The Biology of Glasshouse Pests and Their Natural Enemies*, Koppert B.V., Berkel en Rodenrijs, The Netherlands, p. 109.

Marlow, D.H. 1993. *Greenhouse Crops in North America: A Practical Guide to Stonewool Culture*, Grodania A/S, 415 Industrial Dr., Milton, Ontario, Canada.

Mason, J. 1990. *Commercial Hydroponics*, Kangaroo Press, Kenthurst, NSW, Australia.

Mason, J. 2000. *Commercial Hydroponics: How to Grow 86 Different Plants in Hydroponics*, Simon & Schuster, Australia.

Morgan, L. 1999. *Hydroponic Lettuce Production*, Casper Publications, NSW, Australia.

Morgan, L. 2005. *Fresh Culinary Herb Production*, John Wiley & Sons, Australia.

Morgan, L. 2006. *Hydroponic Strawberry Production: A Technical Guide to the Hydroponic Production of Strawberries*, Suntec (NZ) Ltd. Publications, Tokomaru, New Zealand.

Morgan, L. 2008. *Hydroponic Tomato Crop Production*, Suntec (NZ) Ltd. Publications, Tokomaru, New Zealand.

Morgan, L. and S. Lennard. 2000. *Hydroponic Capsicum Production*, Casper Publications, NSW, Australia.

Muckle, M.E. 1982. *Basic Hydroponics*, Growers Press, Princeton, BC.

Muckle, M.E. 1998. *Hydroponic Nutrients—Easy Ways to Make Your Own*, rev. ed., Growers Books, Princeton, BC.

Nelson, P.V. 1998. *Greenhouse Operation and Management*, 5th ed., Prentice-Hall, Upper Saddle River, NJ.

Nicholls, R.E. 1990. *Beginning Hydroponics: Soilless Gardening: A Beginner's Guide to Growing Vegetables, House Plants, Flowers, and Herbs without Soil*, Running Press, Philadelphia, PA.

Pranis, E. and J. Hendry. 1995. *Exploring Classroom Hydroponics*, National Gardening Association, Inc., South Burlington, VT.

Resh, H.M. 1990. *Hydroponic Home Food Gardens*, Woodbridge Press, Santa Barbara, CA.

Resh, H.M. 1993. *Hydroponic Tomatoes for the Home Gardener*, Woodbridge Press, Santa Barbara, CA.

Resh, H.M. 1998. *Hydroponics: Questions & Answers—For Successful Growing*, CRC Press, Boca Raton, FL.

Resh, H.M. 2002. *Hydroponic Food Production*, 6th ed., CRC Press, Boca Raton, FL.

Resh, H.M. 2003. *Hobby Hydroponics*, CRC Press, Boca Raton, FL.

Roberto, K. 2003. *How-to Hydroponics*, 4th ed., Futuregarden Press, Farmingdale, NY.

Romer, J. 2000. *Hydroponic Crop Production*, Simon & Shuster, Australia.

Ross, J. 1998. *Hydroponic Tomato Production: A Practical Guide to Growing Tomatoes in Containers*, Casper Publications, Narabeen, Australia.

Schwarz, M. 1968. *Guide to Commercial Hydroponics*, Israel University Press, Jerusalem, Israel.

Schwarz, M. 1995. *Soilless Culture Management. Advanced Series in Agriculture*, Vol. 24, Springer-Verlag, Berlin.

Smith, D.L. 1987. *Rockwool in Horticulture*, Grower Books, London.

Straver, W.A. 1993. *Growing European Seedless Cucumbers*, Ministry of Agriculture and Food, Parliament Buildings, Toronto, Ontario, Canada, Factsheet, Order No. 83-006, AGDEX 292/21.

Sundstrom, A.C. 1989. *Simple Hydroponics for Australian and New Zealand Gardeners*, 3rd ed., Viking O'Neil, South Yarra, VIC, Australia.

Taylor, J.D. 1983. *Grow More Nutritious Vegetables without Soil: New Organic Method of Hydroponics*, Parkside Press, Santa Ana, CA.

Van Patten, G.F. 1990. *Gardening: The Rockwool Book*, Van Patten Publishing, Portland, OR.

Van Patten, G.F. 2004. *Hydroponic Basics*, Van Patten Publishing, Portland, OR.

Van Patten, G. F. 2007. *Gardening Indoors with Soil and Hydroponics*, 5th ed., Van Patten Publishing, Portland, OR.

CLOSING COMMENTS

I wish to share with you some comments received over the years from hydroponic enthusiasts. I became involved in hydroponics while a graduate student at the University of British Columbia in Vancouver, Canada. While the initial work was in commercial hydroponic greenhouses, it soon became evident to me that there was a great potential for this type of growing at the hobby level. An engineer and I incorporated a company to build backyard hydroponic greenhouses. We used to go to exhibitions to display our hobby greenhouses and to our surprise found a constant line at our exhibit to enter our greenhouse and see tomatoes, cucumbers, and lettuce growing in rocks. Many people would come in disbelief that these plants were real and they would have to feel them before becoming convinced that they were not plastic. That was back in 1975 when hydroponics was known in the commercial greenhouse industry but not as a hobby.

We began manufacturing small greenhouses of a basic size of 10½ × 12 feet and then larger ones 10½ × 16 ft as shown in Figure 8.1. Most clients thought they were fairly large and would not need all the space, but within months once they were producing their own vegetables they became very diversified in their crops, and of course there were always those houseplants that needed rejuvenating in the greenhouse. Soon their greenhouse was so full they ran out of space. I always remember the biggest complaint we would soon get was "I wish you would have sold me a larger one." The same is true of small indoor units.

FIGURE 8.1 Small 10½ × 16 foot backyard greenhouse.

Once you get fresh vegetables in the middle of the winter, you too will say, "I must expand this hydroponic garden." Our clients found the greenhouse was a very relaxing place to escape from their everyday stresses. They could enter the greenhouse in the middle of a dark rainy day in January and be in a tropical paradise with their plants growing under artificial lights. So there are many benefits of hydroponic growing from pleasure of working with your plants, escape from stress, to a rewarding experience with garden fresh tomatoes, cucumbers, lettuce, and herbs in the height of winter.

My purpose with this book is to show you the various types of hydroponic indoor garden units that are prefabricated and easy to set up for immediate operation. I realize that all such units cannot be covered in a single book, so the aim is to provide you with a cross-section of typical units available under the various hydroponic cultural methods. Most units are modular, so can be fairly easily expanded to meet your increasing demands for more fresh vegetables. There is a lot of help and new ideas, which can be exchanged with fellow growers and manufacturers and their distributors by personal visits to stores or by the Internet. Hydroponic societies disseminate information through conferences, trade fairs, bulletins, or newsletters as well as by the Internet. You should feel free to become part of these organizations and electronic communication sites such as, Facebook, YouTube, and Twitter. Everyone has similar interests and is always looking for new methods and ideas to further advance the science of hydroponics. Enjoy yourself with the plant experience!

Index

Made in the USA
Coppell, TX
12 August 2021